# Competitive Tendering
FOR
# Engineering Contracts

# Competitive Tendering
## FOR
# Engineering Contracts

*M. O'C. Horgan*

OBE, TD, DL, MSc, CEng, FIEE

*London   New York*

E. & F. N. SPON

*First published 1984 by*
*E. & F. N. Spon Ltd*
*11 New Fetter Lane, London EC4P 4EE*
*Published in the USA by*
*E. & F. N. Spon*
*733 Third Avenue, New York NY10017*

© *1984 M. O'C. Horgan*

*Typeset by Scarborough Typesetting Services*
*and printed in Great Britain by*
*J. W. Arrowsmith Ltd, Bristol*

ISBN 0 419 11630 3

**British Library Cataloguing in Publication Data**
Horgan, M. O'C.
    Competitive tendering for engineering contracts.
    1. Engineering—Contracts and specifications
    I. Title
    620      TA180

    ISBN 0-419-11630-3

**Library of Congress Cataloging in Publication Data**
Horgan, M. O'C. (Michael O'Connor), 1910–
    Competitive tendering for engineering contracts.

    Includes index.
    1. Engineering—Contracts and specifications—Great
Britain.   2. Contracts, Letting of—Great Britain.
I. Title.
KD1641.H67   1984      343.42'078624      84-14494
ISBN 0-419-11630-3      344.20378624

# Contents

# Introduction

In the engineering world there are several ways in which two parties can get together in a contractual relationship. Two friends discover that each has something the other needs, so they drift into a mutual assistance pact: a director knows the very firm who are 'experts in that line' so he negotiates a deal with them to develop and provide him with what he wants. But by far the majority of contracts, large and small, involving engineering activities, arise from invited competitive tenders, that is, the would-be purchaser puts down in writing just what it is he wants, and invites suitable firms to make competitive offers for his order. It is with this procedure that we are concerned in this book.

The scope of such contracts is very wide, from a simple purchase of a standard article at one end, to multi-million-pound projects at the other, the size and complexity of the contract matter varying accordingly. It is a matter of some concern how frequently such contracts (be they small, large, or 'in the middle') run into trouble solely through lack of proper attention to vital commercial matters during the tendering period. Anxiety to land the order or to get the technical aspects properly tied up, so often leads to matters of commercial or contractual importance, or prudence, being ignored or overlooked.

The present work is designed to provide for those who often face having to set up engineering contracts a comprehensive text-book and source of ready reference, by the aid of which they may handle the competitive tendering process without doubt or difficulty. Each step is described in chronological sequence, with comment on the common pitfalls. Comprehensive check-lists are included with a detailed alphabetical index to enable a subject to be readily located. As a text-book it aims to be equally of value to engineers and to others who have recently changed from pure technology to new responsibilities with a more commercial, contractual or managerial flavour.

The subject is not, of course, quite so simple as that. We have to be specific enough to provide real practical guidance, yet broad enough not to be circumscribed by any particular organizational arrangement. Especially it is necessary to bear in mind the radically different established ways in which engineers, for the very best of reasons, habitually handle their civil and building construction contracts on the one hand, and their electrical, mechanical and plant-acquisition contracts on the other. In the former, the customer usually assumes all responsibility for design himself, and sets out in his enquiry precisely what he wants: the contractor undertakes to build just what he is told to do, no more and no less. In the plant industry, on the other hand, the customer wants to make full use of the specialized 'know-how', experience and expertise of the plant supplier, so he limits his specification to its functional requirements, and expects the manufacturer to propose in his tender, the detailed design of plant he considers most appropriate. These conflicting approaches are so fundamental to the procedures to be followed that it is worth, at this point, taking a rather closer look at just what is entailed.

## CIVIL, BUILDING AND CONSTRUCTIONAL CONTRACTS

1. Customer responsible for design of the work.
2. Enquiry specification, bills of quantities and drawings give all details.
3. Work is mostly on the site of the customer.
4. Variations to specification usually common and often frequent.
5. Contract price determined by re-measurement and re-pricing of work done.
6. Itemized breakdown of prices and rates in bills of quantities.
7. Frequent payments of contract price based on work done – often paid monthly.
8. No tests on work, or on full completion. Sometimes check-tests on materials used.
9. No documentation, spare parts or customer-training usually called for.

## PLANT SUPPLY CONTRACTS

1. The contractor, in his offer, is responsible for design of the plant.
2. Enquiry specification gives functional requirements only, and ruling dimensions.
3. Work is mostly in contractor's factory: only erection and test of finished plant at site.

4. Variations rare once factory work has started.
5. No re-measurement adjustments. Contract price is firm.
6. Price breakdown unusual beyond major divisions or sections. No bills of quantities.
7. Payment limited to two or three progress payments as defined stages in the work are reached.
8. No materials check usual. Tests on completion, on acceptance, and of achieved performance are essential.
9. Full documentation, stock of spare parts usually called for. Customer-training frequently.

Each approach involves its own tendering routine, and it has been necessary to deal with both in the procedures we set out in this book, without stressing either at the expense of the other.

As might be expected, engineering customers usually have in existence (and well-established at that) one of a host of organizational arrangements for the administration of their contract work. As far as possible, the procedures described have been made sufficiently flexible for the staff-duties involved to be transferred, without too much difficulty, to some other functionary in the firm's existing organization. A warning must be given, however, of the dangers of spreading one individual's duties among several members of an existing staff: each duty carries with it a responsibility, and splitting this several ways means that liability for failure can never be brought home to anybody. That way trouble lies.

To bring into play the greatest number of features, the text has been fitted to one of the larger and more involved types of contract, such, for example, as might form part of a major project. The maximum number of characters is introduced, each with a role to play, including a firm of consulting engineers to act as Engineer for the project on behalf of the purchaser, their client. With the smaller contracts, some aspects we deal with may not arise, or become of relatively small importance: clearly a simpler organization would then suffice, the customer shouldering some of the less-specialized activities with his regular staff. It must be a matter of careful foresight what modifications to and omissions from the procedure can be tolerated without loss of effectiveness, but the actual decision as to just what can be so modified or omitted can only be left to the experience, professionalism, and sound judgement of the reader facing the problem.

As the title indicates, the present work is primarily devoted to direct competitive tendering using single- and two-stage procedures. It will apply equally forcibly to other forms of tender based on competitive offers, such as, for example, the more elaborate versions of 'target-cost' contracting. Indeed many parts of the book will be found of considerable value when calling for and evaluating offers from contractors when no competitive element is involved, as, for example, in negotiated contracts, fixed-fee contracts or 'cost-plus' contracts.

Whenever possible the book is written to embrace operations and contracts outside the bounds of the United Kingdom, both in cases in which an English contract includes work on sites overseas, and those in which one or both of the contracting parties are themselves abroad and subject to foreign laws and customs. It has been necessary to base the procedures described on English Law and customs, and they will therefore be most readily applicable in those countries where the legal system and commercial practices most closely resemble the English ones. Some adjustment of detail will be involved in applying them to contracts under other legal codes, such as the Code Napoleon, or those based on the tenets of religions such as Islam or Buddhism, but as the objectives are the same in all cases – setting up a contract which is sound, equitable and effective – much of what is said in what follows will remain equally true everywhere.

Finally, it should be appreciated the large part that good negotiating can play in establishing good relationships and worthwhile contracts. However carefully an enquiry calling for competitive tenders is prepared, resulting offers will not conform entirely and compromises have to be sought. However, negotiation is an art rather than a procedure and, as such, relies both on flair and experience. It certainly cannot be learnt from a book, and so the coverage given to it here has been restricted to matters of preparation and organization and control of the proceedings. The psychological jungle of negotiating techniques has been avoided altogether.

Whilst every effort has been made to achieve accuracy in compiling this book, no liability is accepted in respect of any errors in, or omissions from, the contents. The names used in the Appendices for the fictitious consulting engineers are themselves fictitious and do not relate to any actual persons. Readers will doubtless appreciate the allusions in the names chosen.

# 1  Contracts, agreements and the Engineer

This book is devoted to the task of setting up lawful engineering contracts between a customer (whom we shall hereafter refer to as 'the Employer' – a title frequently adopted in contracts themselves) and a contractor, the latter selected because his tender has been judged the most attractive of the competitive offers made in response to an enquiry by the Employer. Before doing so, however, we need to establish certain facts about what is necessary to constitute a binding contract, and to say something about the duties of 'the Engineer' (an official usually nominated in engineering contracts) and some of the other officials who support him in his duties.

## 1.1  A CONTRACT

A *contract* is a bargain agreed between two (or more) parties, for example between an Employer and a Contractor regarding the execution of certain works. When properly set up, it is legally binding upon them, each to perform the various obligations he has undertaken, as expressed in a mutually agreed set of *contract documents*. To be legal, a contract must fulfil certain requirements.

- *An intention*, that is to say, both parties must wish and intend their bargain to be enforceable at law.
- *A genuine consent.* The bargain must not be procured by force, coercion or undue influence, nor must it rest on fraud or misrepresentation by one of the parties.
- *Legality* of the object, i.e. the parties must not agree to break the law, for example, to avoid the Restrictive Trade Practices Act, or the payment of Value Added Tax when it is due.
- *Legal capacity of the parties to act.* This rule deals with certain classes of persons (e.g. infants, lunatics, etc.) which do not usually affect engineering contracts. There is, however, a requirement to ensure that

1

persons committing a corporate body (such as a contracting firm) are properly entitled by their constitution so to do.
- *Valuable consideration* must pass both ways between the parties (namely, a bargain as opposed to a gift). In engineering contracts this is usually money for goods or services, but is not necessarily so. The consideration can be a right or benefit, or equally a loss, forbearance or detriment of value accepted by the other party.

## 1.2   OFFER AND ACCEPTANCE

A contract is formed when an offer (such as a tender) by one party is unconditionally accepted by the second party to whom it has been made. Note that the acceptance must be full and complete: any 'ifs' or 'provided thats', or the introduction of new factors, will not form a contract. In the eyes of the law, such an 'acceptance' is in fact a counter-offer which in turn needs unconditional acceptance by the other party. The acceptance must be made within a reasonable time of the offer, and the limit is usually set by the tenderer specifying a 'validity period' on his offer. Acceptances must be communicated to the tenderer – silence cannot be assumed to mean consent. This does not mean that the offer and the acceptance have to be in writing necessarily – for most contracts, and certainly most of those in the engineering field, they can be verbal, and this must be borne in mind by any enthusiastic salesmen anxious to clinch a deal. They must beware promising anything or undertaking something which they cannot fulfil: they might find themselves 'stuck with it'. In practice, of course, engineering contracts of the size we are considering are usually written, either in a formal document or in a letter with appropriate addenda. Otherwise it would be impossible to record the mass of technical and contractual detail. One cannot rely, over an extended period, on somebody's early recollections.

There are, however, a few types of contracts which the law requires to be in writing, either completely or in summary. Without going into too much detail we should note those which may occur in engineering work.

(a)   Completely in writing: include bills of exchange, marine insurance, and some hire-purchase contracts.
(b)   Evidenced in writing: that is, memoranda of all the material terms, duly signed. Includes contracts of guarantee, contracts regarding lending of money, and those which relate to land (purchase, leasing, rights-of-way, access, wayleaves, etc.)

A proper appreciation of the constituents of an acceptance and a counter-offer is most important, and often needs careful thinking through before reaching the true answer. It may help to look at one example before leaving the subject.

An engineer, glancing through a magazine, notices an advertisement for a machine he needs, at a reasonable price, so he writes to the advertiser: 'Please supply one machine as offered in your advertisement, at the price stated.' This appears to be an unconditional acceptance – has he established a contract? The answer is: 'No'. An advertisement (like a display of goods in a shop window, even with prices on them) is legally not a valid offer made to our engineer, 'party-to-party'. It is a general invitation to anyone interested to do business with the seller by opening negotiations. This our engineer has done; he has made an offer to purchase on specified terms.

Suppose the seller now replies by letter saying: 'Thank you for your order, which we are pleased to accept, subject to our usual conditions of sale. Delivery will be 20 weeks.' Is there now a contract? The answer is still: 'No'. The acceptance letter is not unconditional: it has introduced two new factors which the engineer did not offer, (a) the seller's conditions of sale, and (b) a specific delivery date. This acceptance is another counter-offer. The engineer now replies: 'Your conditions of sale are fully acceptable to me, as is the delivery you promise. Payment will be made to you direct by our firm's cheque within 30 days of receipt of the machine, as it is not our practice to hand so much cash to the van-driver, as mentioned in your conditions.'

No, we are still not there! Although professing to accept both new items unconditionally, the engineer immediately shows he has not agreed to the terms of payment, so we have yet another counter-offer. The seller now replies: 'We appreciate your caution and agree to your suggestion to pay us by cheque.' At last! But all the correspondence will have to be included in the contract documentation to establish exactly what has finally been agreed.

## 1.3 DEFINITIONS

An essential feature of a contract is, understandably, that the contract documents must set out the terms of the bargain in a legally unambiguous way, and to this end they usually start with a list of definitions. A word must mean the same thing to both parties. A typical set of defined words is given in Appendix 1. Frequently a word is given a special meaning, more restricted than – or different from – its recognized dictionary definition: to draw distinction in such cases, it is written with a capital initial letter whenever it is used in its special sense. Thus, the 'Employer' – the party to the contract – is distinguishable from an 'employer', some hirer of labour or user of something. We shall do the same in this volume whenever it is necessary to draw such a distinction, and a reference to Appendix 1 will then show the special meaning we wish to convey. Apart from 'Employer' the two most widely used terms are 'Engineer' and 'Agreement', both of which warrant a few words of explanation. One or two others will turn up as we go along.

### 1.3.1  An Agreement

All contracts, however made, require agreement between the parties in order to be valid, and as a result, a contract is often referred to as 'an agreement', whatever its actual form may be. However, a contract can, and frequently is, either drawn up or is confirmed in a formal document, couched in the terms and format of a legal Agreement, and concluded under hand or seal. Such a document is itself correctly called an agreement and so, to avoid confusion in this book with the general use of the word, we shall always describe such a document with a capital letter, i.e. 'Agreement'. There is, however, a further complication as it is not uncommon in some overseas countries to draw up all contracts of any size in the form of an Agreement but to refer to them as 'the Contract'! We discuss this subject further in Section 9.2.

### 1.3.2  The Engineer

As distinct from a specialist in technical engineering matters, the Engineer in a contract is an authority appointed by the Employer to supervise and control the execution of the contract by the contractor. The Engineer is not, himself, a party to the contract, and his appointment, authority, general powers and responsibilities are limited to those set out in the various clauses of the conditions of contract as accepted by the contractor. Within the limits set thereby, the Employer may specify his duties more precisely, but he cannot extend them without the written consent and agreement of the contractor.

The Engineer is frequently identified by name in the contract definitions, but when a firm of consultants is appointed to the role, they will be named as a firm, taking corporate responsibility. As soon as the contract has been let, they will name a senior member of their organization as Project Manager, personally empowered to act as the Engineer on their behalf.

Although primarily the Engineer 'runs' the contract on behalf of the Employer, it is now recognized by usage and established in case-law that the terms of the contract may require him to also act as mediator or quasi-arbitrator between the contracting parties, and this is frequently done. Many standard forms of conditions of contract spell out such responsibilities. In making such decisions the Engineer must, in spite of his primary obligations to the Employer, always be fair and impartial (and, moreover, be seen to be such) using the recognized practices and ethics of his profession in any case where the law itself is not an applicable guide. In this dual capacity the independent status and reputation of a firm of consulting engineers have a lot to commend them even though the firm owes a simultaneous duty to their client, the Employer, to watch his best interests. A few moments consideration will show that, in such circumstances, the Engineer's two interests are not nearly so conflicting as might at first sight appear.

In practice, however, the Engineer is not always so independent. Some bodies, especially the larger government-controlled organizations, public authorities and departments, are apt to nominate as the Engineer one of their own staff, with themselves both his employer and Employer. Whilst this situation is usually accepted, *faute de mieux*, by contractors, there must always be some doubt (rightly or wrongly) as to how faithfully the Engineer can, in such circumstances, fulfil his role as arbitrator when the conflicting views involve his own employer.

We have already mentioned that the powers of the Engineer may not exceed those given to him by the terms of the contract. Occasionally the opposite occurs, and the Employer restricts the Engineer's powers without the contractor (or the tenderer, at the tender stage) being made aware of it. Indeed the Engineer may be prevented from fulfilling completely his role of quasi-arbitrator as provided by the contract conditions, to the detriment of the contractor. Such a practice by an Employer is reprehensible and amounts almost to breach of contract, or fraud.

## 1.4 PROJECT MANAGEMENT

There has, in recent years, been a growing tendency world-wide for Employers to accept they do not have the staff resources themselves, and to place the complete handling of their engineering projects in the care of an independent professional Project Management team. To this end they enter into Project Management Contracts with organizations specializing in such work which can provide a unified service to protect the Employer's interests contractually, legally, financially and operationally.

Among numerous 'definitions' of Project Management, it has been quoted as being:

> 'The overall planning, control and coordination of a Project from inception to completion, aimed at meeting an Employer's requirements, and ensuring completion on time, within budget and to the desired standard of quality and performance.'

The concept is not basically new: with slightly modified titles it will be seen to fit exactly the traditional role we have allotted to the Engineer in this book (Section 1.3), namely an independent firm of consulting engineers who manage a project for the Employer, as well as fill the role of Engineer for the contract works. Alongside such an arrangement it has not been uncommon (especially with government-controlled or nationalized industries) for Employers, wanting to retain the day-to-day control of the works in their own hands, to appoint a senior member of their own staff as the Engineer, even though he does not have the ability or the back-up staff to 'manage' the project in all its aspects. They therefore enter a Management Contract with a firm of consulting engineers to do it for them.

However, it is in the building and construction fields that the need has been found most acutely. In many conditions of contract the appointment of an Engineer is not envisaged, and his role is allotted to the Architect who has been professionally involved in the design, to supervise its execution. Few architectural practices have been organized or staffed to control all aspects of a project, and the need for a separate project management becomes all too evident. It is a situation being experienced world-wide, and the solution is making headway in spite of some reluctance by existing authorities. It has been common practice on public building projects in the USA for many years (though usually under other names, such as 'Construction Programme Management' – CPM) and is now extending into the private sector. It is also common in those other countries which have accepted and practise US methods.

Although we shall not be referring to Management Contracts (in so many words) during what follows, their operation can be clearly seen from the activities of the Engineer and his staff which we shall be following in detail. Indeed, they are executing a Management Contract for their client, the Employer.

## 1.5   THE PROJECT MANAGEMENT TEAM

### 1.5.1   The Engineer's representative

The Engineer is usually permitted by the conditions of a contract to delegate his powers (or some of them) and to appoint an *Engineer's representative* and/or other assistants with more restricted powers and authority. The Engineer is usually required to nominate such persons and define their duties by a letter to the contractor after the contract is signed. The Engineer retains sole responsibility to the contractor and the Employer for their activities.

### 1.5.2 The Engineer's staff

In a large project, such as we are assuming as the basis of the book, a firm of consulting engineers, acting as Engineer, will normally appoint the following key personnel, each of whom may in turn call on assistants or advisers in carrying out the duties for which he has been appointed. Naturally (depending on the nature and size of the Employer's project) some of the appointments may be omitted or may be merged with others, as will best fit in with the internal organization of the firm, the facilities available to it, and possibly the terms of its consultancy contract with its client, the Employer.

### The Project Manager

The leader of the Engineer's team, responsible to management for all aspects of the Engineer's role. He constitutes the formal link with the Employer on all matters concerning the execution of the project.

### The Resident Engineer

The representative of the project manager on the site of the work, to take charge of all the detailed site responsibilities of the Engineer. The resident engineer is frequently nominated as the *Engineer's Representative* when such an official is recognized by the conditions of the contract.

### The Project Engineer

An engineer responsible to the project manager for all the technical aspects of a project. In a multi-disciplined project, separate project engineers may be appointed, one for each discipline involved.

### The Contracts Engineer

An engineer experienced in commercial and contractual matters, available to assist the project manager on such questions. He naturally plays a key role in establishing contracts by competitive tender.

### The Quantity Surveyor

Responsible at the enquiry stage of a contract for taking off the bills of quantities and also for evaluating them, to give a pre-tender estimate of the possible cost of the contract. In tender appraisal the quantity surveyor makes the detailed analysis of and commentary on the rates on the bills of quantities and the validity of the tender prices put forward by tenderers.

During the execution of the contract he assists the Engineer in the measurement, recording and evaluation of work progress and the costing of authorized variations. He maintains a constant watch on the comparison of actual costs with the pre-contract estimate.

For effective control of a contract by the Engineer, the quantity surveyor should either be a member of the Engineer's staff or an independent organization appointed by the Engineer and reporting directly to him. The alternative, which sometimes occurs, in which the Employer himself appoints an independent firm of quantity surveyors, reporting directly to the Employer, is much more difficult to operate effectively. Not only is it necessary to institute a careful liaison and system of cross-communication between the quantity surveyor and the Engineer (and his staff), but occasions and subjects can arise in which conflicting views may be given to the Employer by the Engineer and by the quantity surveyor without there having been a chance of reconciling them. Quite apart from the confusion this causes, the Engineer's authority as manager of the project on behalf of the Employer is undermined, whether such is intended or not.

# 2 *Preparatory arrangements*

To ensure that a common procedure for handling tender enquiries and establishing contracts runs smoothly throughout the course of the project, it is necessary to take certain preparatory steps in good time. Besides establishing and promulgating just what the procedure is to be, it is also necessary to put in hand certain time-consuming actions in order to have them finalized before the tender period begins. They include:

- Define and co-ordinate the respective roles of the Employer and Engineer.
- Obtain government, Town and Country Planning, etc., approval as necessary, including financial credits and transactions with Employers overseas.
- Establish the structure of the Engineer's staff, obtain Employer's approval and allocate roles to each member.
- Establish a firm communication plan between Employer–Engineer–tenderer (contractor)–any other official appointed directly by the Employer (e.g. quantity surveyor, architect, landscape designer, geologist, etc.).
- Estimate budget prices for each proposed contract in the project. These will usually need to be updated with improved accuracy from time to time as the project plans crystallize.
- Draw up an agreed contract plan for the project.
- Draw up an agreed project time-table showing programme for letting contracts.
- Assemble or extend an up-to-date register of suitable contractors.
- Check the financial, commercial and technical status of any new contractors placed on the register.

The arrangements for the necessary capital finance, buyers credits or loans must clearly be finalized by the Employer before any action on the project can start, but this is outside the scope of this present book, and is

9

only mentioned as a reminder of the early part it must play in the planning.

## 2.1  CO-ORDINATION OF ACTIVITIES BETWEEN THE EMPLOYER AND THE ENGINEER

### 2.1.1  Establishment of general responsibility

The relationship between the engineering consultant and his client will have been established in a consultancy agreement in which certain terms of reference will have been given to the consultant. The extent of these will depend very much on the purpose of the original retention of the consultant: was it, for example, envisaged at the outset that the consultant was to become the Engineer or was he retained for purely technical assistance in the early development of the project design? Whatever may have been the circumstances it will be necessary to examine the terms of the agreement as soon as it is made clear that the consultant is to be appointed Engineer for the project and is to take a large share of its management on behalf of the Employer. Among other things, a clear division of spheres of action between him and the Employer must be decided and incorporated in a written brief, which must leave no doubt as to the limits of the Engineer's responsibilities. It must be specific as to the extent the Engineer may act in the name of the Employer on his own authority, especially when this involves increased expenditure of funds. The Engineer's terms of reference should cover his responsibilities in relation to:

- Obtaining government and local authority planning consent.
- Project programming and planning.
- Technical design and specification of the project.
- Liaison with outside bodies, British Rail, DoE, MoT, etc.
- Project administration and documentation.
- Cost estimates for contracts/project.
- Cost control of contracts/project.
- Contractual matters affecting the project.
- Variations to specification.
- Management and supervision of the project on site.
- Acceptance and certification in respect of the Works.

The authority of the Engineer is only imposed on a contractor by the terms of the contract. Many standard forms of conditions of contract already allocate responsibilities to the Engineer so that if the Employer wishes the Engineer to act for him to a different extent than is specified in them (or if no such responsibilities are specified), he must amend the conditions accordingly, at the tender enquiry stage, for eventual incorporation into the actual contract.

A similar consideration applies to the tender enquiry period during which there is no contractual relationship with the tenderer. It may be necessary (or at least advisable) for the Engineer to have documentary authority from the Employer setting out the extent to which he is to represent him in handling the enquiry and negotiations arising from it. This is of especial importance when dealing with contracts overseas, and is essential if the Engineer is to act as *agent* for the Employer, committing him to expenditure and contractual obligations.

### 2.1.2 Responsibilities of the Engineer in regard to tender procedure

As a large project will involve a series of contracts, the functions of Employer and Engineer regarding tender procedure must be defined at the outset of the project, so that a recognized routine can be established and used throughout. They must settle their respective roles in such activities as:

- Drawing up the contract plan.
- Drawing up preliminary lists of tenderers, issuing pre-qualification enquiries, assessing replies, and preparing the final list of tenderers.
- The activities of the quantity surveyor, particularly if appointed independently by the Employer.
- Producing budget/pre-contract estimates for each contract.
- Producing the various enquiry documents (see Chapter 4).
- Assembling and issuing enquiries.
- Acting as link with tenderers.
- Issuing amendments to enquiries.
- Receiving and opening tenders.
- Making technical/contractual/financial appraisals of tenders.
- Assembling the complete appraisal document.
- Negotiating outstanding points with the preferred tenderers on technical/contractual/financial matters.
- Issuing rejection notices to unsuccessful tenderers.

Letters of acceptance should always be sent by the Employer even though drafted by the Engineer, as the contract has to be established between the contractor and the Employer. Nevertheless it sometimes occurs that the Employer expects the Engineer to issue a letter of acceptance on his behalf and in this event the Engineer must take some steps and observe extra precautions to establish formally his role of agent acting on behalf of the Employer. He must avoid shouldering responsibilities under the contract which are actually the Employer's.

- He should have his appointment as agent recorded in writing by the Employer.
- He should have written instructions from the Employer as to which tender he is to accept, what negotiated amendments to include, the contract price, and any other matters not covered by the contract documents.
- He must make it clear in any document he signs that he does so as the duly authorized agent for the Employer.
- He must not, in such document, undertake any liability or responsibility himself.
- If possible, he should get the Employer to approve the final draft of a document he is about to sign as agent.
- He must not, as Engineer or as agent execute any deed under seal for the Employer (obviously he would be using his own seal – not the Employer's). If this need should ever arise, he would require powers of attorney, and need to take special precautions to avoid assuming unwanted responsibilities himself.

This may all sound rather involved, and much easier for the Employer to sign his own acceptance: however, occasions do arise, mostly overseas, when contracts are concluded at a local meeting with the other party. The Engineer must always bear in mind that it is the Employer's money he is committing on a contract to produce something precisely as the Employer wants it. If he misinterprets the latter, or overspends the former, he may well find himself called to account for his mistake.

### 2.1.3 Employer's approval

The need for the Engineer to obtain the Employer's approval applies to much more than accepting tenders. Although the former is frequently given blanket authority to commit the Employer as he deems necessary up to a stated limit of spend, it is usually as well even within that limit, first to obtain the Employer's approval, except in very minor expenditure. The same principle should apply to decisions relating to commercial or contractual matters. It is the Employer's contract and money, and he almost certainly has established routines for recording changes and allotting and accounting for all expenditure.

During the preparation of the enquiry and the tendering period, the Employer must go along with what the Engineer is doing, step-by-step, to avoid any possibility of a disagreement on policy appearing later and causing delays and changes in the Engineer's tactics. Approval should therefore be sought for:

- The list of contractors to be included in the preliminary enquiry.
- The final list of tenderers for the main enquiry.
- Each document forming the tender documents included in the main enquiry.
- The instructions to tenderers issued with the enquiry.
- The date and place for the return of tenders, and any changes thereto.
- Proposed visits by tenderers to Employer's site.
- All amendments to tender documents, prior to issue.
- Tender reception and security arrangements.
- The form the appraisal reports are to take, and the extent of detail the Employer expects therein.
- Outstanding matters to be negotiated with tenderers and the aims it is hoped to achieve.
- Any objections by Employer to proposed list of preferred tenderers.

Besides being a question of principle when dealing with a client, the constant seeking of approval in this way is also a matter of prudence. The Engineer can be very embarrassed if he discovers late in the day that he has committed the Employer to a course of action he never intended.

### 2.1.4 · Intercommunication

1. The efficient running of any project depends to a very large extent on a properly organized system of communications covering not only letters but telephones, telex and cable, circulation of drawings, records of meetings and the rest. The basic requirements of the chosen system are:

   (a) All concerned with a piece of information (either for action or as background intelligence affecting their decisions) must receive it promptly and, if possible, first hand.
   (b) Staff must not be bogged down by being sent masses of papers which do not affect them.
   (c) Formal documents affecting basic contract/project decision-making must be channelled through the chain of individuals authorized to take or approve such decisions.
   (d) Communication at technical level in any discipline should be unfettered and not restricted through a specified single point of contact.

(e)  The two main groups involved (the Employer and the Engineer) already have their own routines and methods, so that any selected system should, as far as possible, merge readily with the existing ones.

(f)  Routines selected must be as simple as possible, to allow them to be followed religiously: individuals should automatically know who else is already in receipt of a piece of information, and whom they themselves have to inform.

2.  The methods actually adopted must depend, in detail, on the circumstances of each case, but the following will usually be found worthwhile between the Employer and the Engineer.

(a)  Technical liaison by any means of communication between opposite numbers in any one discipline is permissible provided it is restricted to technical matters.

(b)  For policy matters (including any decisions on technical matters affecting the costs of contracts/the project), a single channel of communication between the Employer and the Engineer is designated for each of the following activities:
  ● Technical (in large projects this may be one channel for each engineering discipline involved).
  ● Contractual.
  ● Financial.
  ● Administrative.
  ● Project management.
  It will be a matter for local decision the points beyond which matters of policy within these five activities have to be channelled through the project manager himself and through his 'command' link with the Employer.

(c)  A common system of references for numbering and identifying documents should be agreed for use by both the Employer and the Engineer. This will usually be based on recognizable initials:
  ● A project reference.
  ● Sub-references for each major division of the project.
  ● Originator's identification (usually by an extension of his existing internal identification).

(d)  A single channel of communication only should be allowed with tenderers. Depending on allocation of responsibilities in Section 2.1.2 above, this may be from the Employer or from the Engineer but not from both. A single official should be named.

(e)  When other organizations are appointed to the project by the Employer and not placed directly under the Engineer (e.g. quantity surveyors, architects), special tripartite liaison arrangements must be made to suit the particular circumstances. They need to be planned with the utmost care to avoid the Engineer's responsibility for project management being undermined, and to obviate the possibility of inconsistent decisions or instructions being given to them by the Employer. It is usually a first essential that the 'command links' from such organizations should be to the same individual on the Employer's staff as is the Engineer's link. Any arrangements made with them by the Employer in regard to the project must automatically be copied to the Engineer's project manager (and indeed should preferably be made only with his prior approval).

(f)  The system should nominate only a limited number of degrees of confidentiality applicable to all forms of communication and must specify for each the security measures to be adopted and the limits of distribution.

3.  *Circulation lists.* The most readily applicable method of ensuring that all concerned always receive documents which are important to them is to establish for the project standard circulation lists. Haphazard, *ad hoc* or 'need to know' methods invariably result in people only receiving some of the communications relating to a given matter: the items missed usually have a habit of being the ones which, in the long run, concern them most!

Routine standard circulation lists avoid this. In complex projects they also simplify the work of the secretarial staff and automatically ensure that all recipients know who else has already been informed. To keep the number of copies to a minimum, a single list is usually replaced by several, each identified by a reference letter. No individual appears on more than one list. The distribution for any document can thus be specified by the list references to which it has been sent, e.g. 'List A' or 'List D' or 'Lists A, B and D', and the document marked accordingly. Both the Employer and the Engineer should have one list showing the other party's link representatives, so that documents can be cross-copied into the distribution system of the other organization.

In two alternative methods, both well recognized, lists can be either of individuals or of departments in the firm's organization. The individual or the heads of the departments can recopy to subordinates inside their own domains, if this seems to them to be required.

*Lists based on individuals in the management structure*
List A  Individual directors concerned with the project.
List B  Individuals at project management level.

List C  Individuals involved with the project at departmental level.
List D  Engineers, cost-controllers, etc. (often divided into sub-lists D.1, D.2, etc.).
List E  Link members of other organizations (e.g. Employer, quantity surveyors, etc.).

*Lists based on departmental strucuture*
List A  Managing Director's Office (includes Company Secretary, Chief Accountant, etc.).
List B  Project Manager's Office.
List C  Contracts Manager's Office.
List D  Technical Departments (often sub-divided into sub-lists for each discipline).
List E  Project Cost Control Department.
List F  Link members of other organizations.

The choice to be adopted between these alternatives will usually depend on the precise nature of the project and the respective duties of the Employer and Engineer, and also the existing organizations and routines in their firms. There are considerable advantages in using the same standard system in both organizations if this can be tolerated.

4.  *Drawings.* Engineering drawings form one of the largest single items of documentation needing distribution. In most cases there is no call for accompanying correspondence (apart from the purpose of circulation). As a result, drawings all too often arrive unannounced in large rolls or bundles with little or no indication of their origin or the purpose for which they have been sent.

It is advisable to adopt for the project a standard routine for recording the movement of all drawings (and, indeed of any other documentation for which a covering letter is not required). This involves the introduction of a formalized issue note on the lines suggested in Appendix 22, a copy of which must accompany every packet of drawings. The form is initiated by the person issuing the drawings, each issue being marked with the identification of the issuing office and its serial number. A copy is sent to all recipients of that issue and a file copy is kept by the issuing department. It will be seen that the form embodies all the basic information usually required in these circumstances.

5.  *Master Contract Record.* We shall see in Chapter 10 the need for ensuring that a copy of every document relating to a contract is kept on a master contract record file. Arrangements for this to be done must feature prominently in any system of intercommunication and circulation lists which is adopted.

## 2.2 THE CONTRACT PLAN AND THE PROJECT PROGRAMME

At the outset of the Employer's project, these two basic planning and control documents must be drawn up. They are completely interdependent and must therefore be considered together.

### 2.2.1 The Contract Plan

This is the proposed division of the project works into suitable contract 'packages', the objective being the lowest project cost related to the shortest overall project time. It must include not only work which it is intended to put out to tender but also any items carried out by direct labour. Factors which need to be considered when manipulating the portions of the works into their optimum contract packages will include:

- The compatibility of the various portions of the works for combining into a single contract: their potential for using common items of plant or specialist labour.
- The intended work sequence as shown by the project programme, especially items on the critical path.
- The amount of work which can be accommodated on site at any one time, or parts of the work which involve the same part of the site.
- Limitations on the project programme imposed by the time required to design portions of the works to a standard suitable for enquiry purposes. Effect of grouping such items with others already designed.
- Availability of construction labour and materials in the area, and time for provisioning (especially in projects overseas).
- The dependence of other inter-related parts of the works on the progress and completion of the proposed package.
- The structure of industry in the area: how readily could it undertake the proposed package as a single contract. Availability of specialist contractors.
- The financial advantage (or otherwise) of grouping the parts of the works in the proposed package.
- The extent of competition among firms to undertake such a package.

Each factor will have a relative degree of importance in the particular circumstances of the project, and with appropriate bias being applied to the pros and cons, a grouping of the works into selected packages will be arrived at. This contract plan is now used to adjust the project programme, and if necessary the procedure repeated, until a satisfactory contract plan consistent with the limitations of the project programme has been arrived at. From this eventual programme the detailed dates affecting each of the constituent contract packages can be deduced:

- Deadline for completion of design.
- Deadline for issue of enquiry.
- Date for return of tenders.
- Contract start date.
- Contract completion date.
- Intermediate dates for completion of any critical sections of the contract works.

In addition, the contract plan can now be expanded to show for each package:

- The exact scope of the package.
- The type of contract proposed (i.e. negotiated, cost-plus, etc.).
- The standard form of conditions of contract appropriate to the works in the package.
- Budget capital cost of the package.
- Proposed nominated sub-contracts (if any).

### 2.2.2 The Project Programme

The Project Programme shows the intended progress of work on the project from initiation until final completion. An outline programme is first built up either as a bar chart or as a network diagram showing the critical path (or both). All the key elements of the project must be shown with the estimated periods they will require. These will include:

- Establishing financial/credit arrangements (especially for overseas projects).
- Land purchase, wayleaves.
- Planning consents.
- Infrastructure, including road, rail, port facilities, availability of services, water, soil investigation, site layout, construction and storage arrangements.
- Preparation of designs and specifications for each portion of the works.
- Procurement (with particular regard to long-delivery items), but allowing for the issue of enquiries and evaluation of tenders for each contract package.
- Construction times and sequences for each part of the works, and each contract package.
- Commissioning and testing of the project works.
- Operational staff training.

The process of setting out the network diagram or the bar chart may well indicate that any contract plan previously proposed is not the most suitable, so that a revision is made until an optimum is reached.

**2.2.3** The reliability of the project programme drawn up at the start of a project depends largely on:

- The degree of finality reached in the project design when the programme is worked out.
- The extent of past experience on similar contract packages and hence the accuracy of the estimates for procurement and construction periods.
- The assumptions made of the load on the industries which are going to be involved, by the time their contracts occur in the project programme.
- The depth of consideration given to the phasing of the construction work.
- The intensity of the load which the contract grouping and the programme itself will place on the Engineer's organization and in particular on the sectors dealing with design, specification and procurement.

The Project Programme (and to some extent the Contract Plan) will clearly be subject to modification as the project develops from its early concepts. They will need constant review throughout the project, not only as the design develops in detail and crystallizes, but also as various unforeseen (and unforeseeable) events affect the actual progress achieved. No two projects are identical and with the best experience in the world and the most careful planning, actual progress on a complex project is unlikely to match the early predictions.

## 2.3 REGISTER OF CONTRACTORS

### 2.3.1 Maintenance of a register of contractors

Much time can be saved when issuing an enquiry if a list of potential tenderers can be assembled without delay from a comprehensive register of contractors of all types, compiled in advance. Most consultants maintain such a register, but its usefulness in a given project must be checked as soon as the contract plan has been settled, in case:

- Existing relevant information needs up-dating.
- Some contract works are of a nature not covered by the current register, in particular specialist equipment.

- The project location (e.g. overseas) may necessitate the use of contractors from areas not covered by the register.
- Longer tender lists are required than the existing entries can provide.

The additional data should be obtained as an advanced preparatory measure, using the information and suggestions given in the following sections. Many firms already on the register will probably be well established, with names and reputations well known in the engineering industry. The combined experience of the Engineer's staff will, in many cases, be able to provide any additional details about them without special investigation being necessary. Otherwise the following sources and methods may be applied.

### 2.3.2   Sources of information regarding contractors

Once the Contract Plan has determined the size and scope of the contracts which need to be placed, the names of possible contractors can be obtained from the following sources and added to the general register:

- The Employer (especially for specialist plant).
- Trade directories.
- Trade associations.
- Professional engineering institutions.
- Technical press and their information departments.
- Commercial attaches of foreign embassies in UK.
- Commercial attaches of British embassies abroad.
- Commercial directories (e.g. Sell's, Kelly's, Kompass, Dun and Bradstreet).
- Information departments of Department of Trade and Industry.
- Information department of Confederation of British Industry.
- Any known previous employers who could be approached without danger of disclosures embarrassing to the Employer.

Some of the above sources have corresponding organizations in foreign countries. The Kompass directories are an excellent source of such information in countries for which they are published. Others (such as the DTI) have departments devoted to information about industry and industrial organizations in foreign countries. And finally, overseas countries which are industrially well-developed will themselves have trade and similar associations from whom details of their industry in that country can be obtained.

### 2.3.3 A standard form of request for information

Data on some promising contractors obtained from the above sources may still lack important features which need to be made good before the Engineer could reasonably recommend them to the Employer for inclusion in any particular list of tenderers. The best approach at this stage is direct to the contractor. A standard form of request for information, similar to that shown at Appendix 3 hereto, can be sent to all seemingly useful contractors by the Engineer, with a letter inviting them to supply the information being sought 'to be included in our records'. Few, if any, contractors will refuse.

To prevent premature disclosure of the Employer's intentions, his identity and the nature of the project will not be revealed at this stage either directly or by inference (for example by too detailed a description of specialized equipment). The contractors will be given no grounds for expecting to be included in the list of tenderers for any particular contract.

### 2.4 INVESTIGATING A CONTRACTOR'S COMMERCIAL AND TECHNICAL SUITABILITY

### 2.4.1 Scope of information

A properly completed standard form of request for information will normally give an adequate picture of a contractor's capabilities, but if the project in view has unusual aspects, the form may need elaborating by the addition of a further question sheet. Thus a foreseeable requirement for handling exceptionally heavy loads, under-water working, unusual security arrangements or suchlike, would provoke further questions. If the Employer, the project site or the contractor is located overseas, contractors should also be asked for information regarding, for example:

- Their familiarity with methods and practices in the country concerned.
- The availability to them of linguists in the required languages/dialects.
- Their knowledge of communications/transport/shipping/customs problems and routines applicable.
- Their knowledge of business/political/religious/social/accommodation/welfare problems in the area.
- Their knowledge of meteorological/climatic/topographical/health and hygiene/natural resources and similar details in the area.
- Experience in working with local/foreign/immigrant labour including camps/staff/diets/work permits/leave arrangements etc.

### 2.4.2   Taking up references

References to previous employers given by the contractor should be followed up. This should normally be done by the contracts engineer on the Engineer's staff who will consult with the project engineer when the references are in respect of technical matters. Wherever possible the referee should be asked to report on specific aspects of a contractor's performance: a general request for information usually produces a generalized, indefinite and rather useless response. A questionnaire to be answered is a good approach. Points of doubt might include, for example, the contractor's record for the following:

- Management/administrative prowess.
- Contract control and organizing ability. Integration of working forces.
- Claims reputation.
- Dependence on sub-contracting.
- Accuracy of work to drawings. Avoidance of wasted work.
- Ability to co-ordinate the use of diverse earth-moving machinery.
- Programme keeping; ability to meet target dates; sense of urgency. Avoidance of excess waiting time.
- History of strikes and labour unrest.
- Special technical features of plant which can be supplied.
- Performance/efficiency/reliability of plant supplied.
- Availability of spare parts/rapid maintenance services.
- Co-operative/obstructionist outlook.
- Technical matters in which the contractor's competence may be in doubt.

## 2.5   INVESTIGATING A CONTRACTOR'S FINANCIAL STATUS

### 2.5.1   The Engineer's duty to the Employer

An Engineer who accepts responsibility from the Employer for handling tendering procedure or managing a project on his behalf has a duty to consider the financial capacity and stability of any tenderer he recommends for the award of a contract. If a contractor fails in mid-contract, the loss to the Employer through additional contract costs and through delay and disorganization of the project can be very high and the chance of recovering any recompense from the contractor is low. He may be tempted to try and recover from the Engineer on grounds of negligence. To what degree is the Engineer liable?

In dealings with their clients, engineering consultants do not usually claim any expertise in the specialized field of company finance and would take care not to allow anything to be written into their consultancy agreements to suggest they do. The same precaution should be applied to the terms of reference they accept on appointment as the Engineer for the project. In such circumstances, the Engineer's duty in this respect would be to take reasonable precautions to ensure that any contractor he recommends will not damage his own reputation with the Employer by failing on the contract. He is neither entitled to ignore the financial status of any contractor as being none of his business, nor is he obliged to make a thorough and professional investigation (as he is expected to do in regards to the engineering aspects). If an Engineer who has disclaimed financial expertise takes care to establish *prima facie* that a contractor is financially sound he will avoid any suggestions or claims of negligence by his client, the Employer, if things later go wrong for financial reasons. Clearly he must reveal any doubts he may have to the Employer for his further investigation.

## 2.5.2 The time factor

Strictly speaking only the firm recommended for the award of a contract needs to be financially vetted. There are, however, advantages to be gained by extending the scope to the whole of the provisional list of tenderers, but the process can take some time (especially in borderline cases) and such a step assumes that the list can be established sufficiently early in the project. The advantages are:

- Doubtful firms can be excluded from the final tender list, making way for more reliable ones.
- Financial status is an important factor during appraisal of tenders, especially if the lowest bidder is of unknown status or appears to be too small for the size of contract.
- The final selection of tenderer can be changed at a late stage, without incurring delay for vetting the new choice.

It will be appreciated that in the case of many contractors of long standing and repute, their reputation can reasonably be accepted by the Engineer without further action: it is generally the small, lesser-known firms which need to be vetted, especially as they are liable to put in the lowest bids for a job.

### 2.5.3   Financial investigation procedure

1. Indications of the credit-worthiness of a firm may be obtained from one or more of these sources:

   - The firm's annual reports, balance sheets, official declarations and returns open to inspection at various public record offices.
   - Statistical reports (such as the various 'financial ratios') published from time to time in business and other journals.
   - News reports, stock exchange movements and the like published in the national and trade press.
   - Credit agencies (which normally provide condensed data from the foregoing sources).

   Some contractors (especially smaller ones) may be able to obtain (at the request of the Engineer) a banker's reference from their bank, but this is not likely to be an acceptable approach unless the actual award of a contract hangs on the result. Apart from this, banks are rarely useful as they normally will not disclose information about their customers' financial position except to other banks.

   A possible source of information may be other recent customers of the contractor who made enquiries on their own account, and might be prepared to give 'off-the-record' comments as a result of their own dealings with the firm in question.

2. Less direct, but none the less useful, information can be obtained from some of the sources listed in Section 2.3.2 above especially on such matters as:

   - Type of company (limited, private, partnership, etc.).
   - Parent company or associated companies.
   - How long established.
   - Issued capital.
   - Average annual turnover.
   - Number of employees maintained.
   - Scope of their business (e.g. does it include capital-greedy activities?).

3. Where closer examination is deemed necessary, a careful and tactful approach to the firm itself can be very revealing to the experienced investigator even though the evidence may be somewhat intangible. An atmosphere of confidence, frank and ready answers to questions about its operations, the use of efficient systems of contract control and administration on its current contracts, no recent history of staff redundancies or stringent economies, no unusual rate of staff resignations, all help to build up the picture.

### 2.5.4 Fully-owned subsidiaries

Special care is necessary with a subsidiary company which has little or no capital in its own right to back its operations. An unsuccessful subsidiary can be jettisoned by its parent to save the expense of a rescue operation. It is therefore common practice before awarding substantial contracts to such subsidiaries to require a performance bond from the parent company. A typical form of such a bond (which must be executed under seal and stamped) is shown at Appendix 29. Alternatively the contract can be offered to the parent company itself with approval for it to sub-contract to its subsidiary but not to assign the contract itself.

### 2.5.5 Assistance to a contractor by the Employer

Under today's conditions few firms can take on a contract of any size for the supply of plant (or, indeed, any other contract not involving regular monthly – or more frequent – payments on remeasurement) without financial assistance, either by interim payments of the contract price or credit loans from elsewhere, to meet their cash-flow demands. In the case of credit-loans the interest they have to pay will, of course, be recovered from the Employer as part of the contract price.

The employer may well have access to cheaper financing arrangements himself, and if so can save money by agreeing to generous interim payments of the contract price. Although this will ease the contractor's cash-flow problem on the Employer's own contract, it is, of course, no guarantee of the continued solvency of the contractor overall. The Engineer would need to obtain a reliable picture of the contractors financial status before recommending the acceptance by the Employer of such terms of payment. See also Section 7.6.2 relating to buyers credits.

## 2.6 INFORMAL ENQUIRIES FOR ESTIMATED PRICES

When estimating prices (for example, for budgets) and especially in the case of specialized plant, it may be necessary to obtain from manufacturers closer estimates than can be ascertained from the Engineer's own internal sources. The same can occur in connection with modifications to more normal or standard specifications for plant which the project may require. In such cases, whether the approach be verbal or written, it must be made manifestly clear to the supplier that it is not an invitation to tender and any figure he gives will not commit him in any future tender.

It is usual and helpful to tell him what tolerance on the accuracy of the estimate can be accepted, and this should naturally be made as wide as possible to avoid his having to do a close costing exercise. The contractor is being asked at this stage to do a favour and he should not be expected to embark on a lot of calculations to arrive at an answer any more precise than is necessary for the Engineer's immediate purposes.

As with other preparatory arrangements, care must be taken to avoid embarrassment to the Employer by premature disclosure of his project plans, and to avoid any commitment to include the contractor in any subsequent tender enquiry. When one is dealing with a specific type and size of machinery, this can frequently prove a matter of some difficulty. Provided that prior approval has been obtained from the Employer, it may be better in such a case to explain the full position in confidence. There is, however, a moral obligation thereafter to include the firm in any subsequent list of tenderers for the plant concerned.

# 3 The enquiry

## 3.1 SELECTING THE LIST OF TENDERERS

Much depends on a good list of tenderers from whom reliable offers and dependable execution of the work can be expected. Considerable care should be taken in drawing it up. However, for every tenderer preparing his offer is an expensive matter, and only one can get the order and recover his costs. If too few bona fide tenders are received from an enquiry the Employer and the Engineer face the expense of repeating the procedure, with a serious loss of planned project time whilst it is being done. The length of the list of tenderers is therefore a compromise between a good spread of offers and a minimum number of out-of-pocket contractors. The number actually chosen depends on the type and extent of the work involved and sometimes on the views of the Employer: as we shall see in Section 3.3, the final list should be between three and six names, and the length of the preliminary list should reflect the expected number of rejections.

The starting point is the register of contractors described in Section 2.3. In conjunction with this are taken the results of any investigations made as to the technical, commercial and financial suitability of contractors, as described in Sections 2.4 and 2.5. In this way a provisional list of tenderers is drawn up and submitted to the Employer for his approval. No firm should be included whose interests are only borderline to the work of the contract: there is too much risk of being 'experimented on' should such a firm submit the lowest offer and be awarded the contract. In the same way, the manufacturers of a range of standard products are not likely to be interested in supplying modified versions unless the quantities are large enough to make the tool changes and disruption of production flow worthwhile.

The Engineer must take full account of the views of the Employer, who can have very firm ideas as to the contractors he favours, and indeed, on those he insists on avoiding. The Engineer himself is not allowed such feelings, which could impair his impartiality: his advice and recommendations must be based only on factual data. On the larger or more complex contracts where a proportion of firms approached may not (for one reason or another) be prepared to make an offer, the initial list is usually made correspondingly longer, and reduced to a reasonable minimum by the *Preliminary Tender Enquiry* procedure which we deal with in Section 3.2. Nevertheless, the final list must allow a margin for firms which, having agreed at the preliminary stage to compete, suffer a change of heart when they examine the full enquiry. If the contract is judged to be an 'unpopular' one, or the particular industry is known to be heavily loaded already, the number of refusals at the preliminary stage can be quite large, and to them must be added one or two non-competitive 'face-saving' offers by firms who consider it bad for their image to refuse to tender, even if they do not want the job. These must both be reflected in the length of the provisional enquiry list.

## 3.2   THE PRELIMINARY TENDER ENQUIRY

**3.2.1**   The Preliminary Tender Enquiry seeks to eliminate from the provisional list of tenderers all who are unsuitable, or are unable, unwilling or incapable of carrying out the contract concerned. It can simplify and cheapen greatly the subsequent effort and expense involved in producing sets of tender documents and appraising or rejecting the resulting offers. In its simplest form it is a letter saying that an enquiry is to be issued shortly and asking if the recipient would be interested in tendering. A typical example is shown at Appendix 4. The letter should always include adequate data to enable the contractor to appreciate the type of work, its location and size, and the expected period during which the contract will run. He is then in a position to consider it carefully in relation to the resources which will be available to him. Some or all of the following points may need to be commented on in the preliminary enquiry.

- The Employer's name and the title of the project.
- The location of the site.
- A short summary of the proposed contract work, bringing out any abnormal or unusual features.
- The extent of any design, development or drawing work expected from the contractor.
- Indications of the magnitude of the work (e.g. dimensions, production rates, power, etc.). Approximate costs or prices should *not* be quoted to indicate size.

- The expected date of issue of the enquiry.
- The intended date for the return of tenders.
- The expected contract programme: date of placing the contract; start date; completion date.

**3.2.2** In the case of large, complex projects and especially those for which tenders are to be sought internationally, the object of the preliminary tender enquiry has become more inquisitive, as befits the greater demands they will place on the resources of the contractor. It seeks to establish for each firm, its suitability for further consideration from each of a number of standpoints, and goes into considerable detail.

- Is the firm of adequate size and status to handle a contract of this magnitude?
- Has it any recent experience of similar work, preferably in similar circumstances?
- Does it have adequate technical expertise?
- Does it have appropriate quantity, size and types of contractors equipment? Or suitable factories?
- What is its financial background and has it adequate financial strength or credit backing for a contract of this value?
- Has it the experience and organization to handle overseas contracts, especially in the country of the current project? Can it handle shipping and export documentation? Has it got translator and interpreter resources?

In such cases, the preliminary tender enquiry assumes large proportions, a lengthy questionnaire rather than a short letter. It is no longer a question of asking the contractor if he wants to tender, but requiring him to satisfy the Engineer that he is capable of undertaking the work. As we shall see in Sections 3.4 and 5.4.2, it may even take the shape of an actual proposal from the firm as to how it would propose to handle the contract if it were given it.

## 3.3   THE FINAL TENDER LIST

Project programmes rarely allow time for repeat tendering, and the preliminary enquiry offers the best insurance against this being necessary by eliminating the non-starters. All recipients who reply affirmatively to the simple preliminary enquiry letter must subsequently be invited to tender. With the more searching enquiry, only those whose replies satisfy the Engineer and the Employer, after a full and careful appraisal, are subsequently invited to tender. In the case of some overseas firms whose replies are, on the face of things, quite in order, it might be necessary for the Engineer to carry out more detailed investigations by taking up references, or using consular resources.

In any case, the preliminary enquiry leads directly to the final tender list, and it is for this reason that the Employer's approval of the preliminary list is required. The final list should never be less than three names plus one extra in case of last-minute refusals. A maximum of six or seven names should normally suffice.

Almost as big a problem as having too few potential offers to form a reasonable list, is having too many! This can readily occur when the advent of a large project or contract receives wide press publicity, especially during a slack period in that branch of the industry. High local and national interest is aroused, and many equally eligible and respected firms write in formally requesting an opportunity to quote.

The Engineer's problems is then to reduce the number to a reasonable list without damaging his (and the Employer's) reputation for impartiality, and laying himself open to charges of favouritism by the unlucky suitors. Clearly a selection based on a succession of logical criteria must be applied. Typically the following might be considered:

- Eliminate firms whose annual turnover is less than £x (a figure deemed appropriate to indicate firms of insufficient size to handle contracts of the value concerned).
- Eliminate firms with no previous experience of a similar project in size and technical requirements.
- Eliminate firms having to sub-contract substantial portions of the work.
- Eliminate firms not having offices or branches in the locality of the site, or in the overseas country concerned.
- Eliminate firms with low issued capital, or who are dependent on an associate or parent firm for their finance or credit.

Finally it may become necessary to apply further pruning of the list by considering 'preference' criteria. Does the list contain firms:

- To whom the Employer has a duty (e.g. sister or associate or subsidiary companies)?
- With whom the Employer does considerable counter-business?
- Who have been previously employed on similar work with notable success?
- Who have a reputation for good labour-relations?
- For whom the Employer has a predilection?

In suggesting these criteria it has been assumed that any firms whose ability, methods, organization or reputation are known to be below standard have already been excluded from the list, leaving only real candidates.

## 3.4  INTERNATIONAL ORGANIZATIONS – PROCUREMENT METHODS

### 3.4.1  Credit organizations

On the international scene are a number of credit agencies, most of which have their own special rules and regulations for tendering and contracting procedures. They include, for example, the World Bank Group, the United Nations Development Agencies, a number of regional agencies such as those operating in the Arab countries and African 'Third World' areas, and a number of others. Their interest in such procedures is obvious: they must ensure their loans are fairly and economically applied for optimum realization of the purposes for which they have been granted. A publication issued by the British Overseas Trade Board entitled *The International Lending Agencies* is recommended for further details.

Typically, they call for two-stage tendering in one of its numerous forms (see Section 5.4.2). The World Bank, for example, requires the Employer to issue a 'Pre-qualification Enquiry' internationally. They require to discuss and approve the Employer's proposed enquiry documents and preliminary list of tenderers, and subsequently, after the bids have been received, the Employer's appraisal of them and his final tender list. Details of this and other procedures relating to tendering against World Bank credits are set out in their booklet *Guidelines for Procurement under World Bank Loans and IDA Credits* published by the International Bank for Reconstruction and Development and The International Development Association. This is a booklet which needs reading carefully, as it tends to state requirements of immediate importance almost casually, easily to be overlooked.

### 3.4.2  Public authority contracts – special EEC* requirements

The EEC has promulgated special regulations in respect of tenders placed by public authorities in Common Market countries for (a) supplies, and (b) civil engineering and building contracts. These regulations aim at

* European Economic Community

abolishing any form of discrimination between tenderers on grounds of their nationality, and they cover:

- Rules for drawing up specifications.
- A requirement to advertise enquiries in the *Journal of the EEC*.
- Matters to be observed in receiving and appraising tenders and awarding a contract.

As the regulations presently stand, they apply only to contracts having an estimated total value: in the case of supplies, 200 000 EEC Units of Account (ECUs), i.e. about £110 000 at present values; and in the case of civil engineering and constructional contracts, one million ECUs (about £550 000): in arriving at this figure, the value of any nominated sub-contracts is ignored. Tenders may *not* be split with the intention of avoiding the regulations.

A 'Public Authority' includes local authorities, all government departments, new-towns corporations, Scottish Special Housing Association, Northern Ireland Housing Executive, etc. However, government-controlled bodies such as British Railways, Atomic Energy Authority, British Steel Corporation, British Airways, etc., are not embraced by the definition.

In addition to the minimum financial limits stated above, there are numerous other circumstances in which the requirements of the EEC directives are relaxed. These cannot all be detailed here, but as an example, they do not apply to the following classes of works:

- Industrial plants of mechanical, electrical or power generation (or distribution) nature.
- Nuclear plants (industrial or scientific).
- Works specifically in connection with mineral extraction (e.g. mines, drifts, tunnels, shafts, quarries, etc.).
- Variations to existing contracts, or repeat contracts.
- A number of specialist exclusions, individually specified.

The Department of the Environment publishes three circulars which give fuller details and also give references to the actual EEC directives which apply. These circulars are:

- In the case of supply contracts No. 46/78.
- In the case of civil engineering and constructional contracts, Nos 4/73 and 59/73.

## 3.5 THE PERIOD NEEDED BY A TENDERER TO PREPARE HIS OFFER

**3.5.1** Tenderers should always be given adequate time to prepare their bids – a rushed tender invites errors and omissions which cause extra work and delay clearing them up later. The tender period (being the last stage of the chain) is all too often squeezed by the cumulative delays which have occurred in drawing up and issuing the enquiry documents, and the responsibility for these usually rests very close to home. The issue by the Engineer of a realistic enquiry programme (see Section 3.6) well in advance, and a determined adherence by all concerned to its dates, are the only good ways of avoiding last-minute rushes.

**3.5.2** Allowance must be made for the fact that tenderers are usually not sitting idle, poised to deal with the enquiry immediately it arrives. In addition, in formulating an offer they may need to get involved in design work, certainly in planning their staff and machine resources, and the like; they may have to produce enquiries for their sub-contractors, get prices from them and delivery promises from their suppliers. Thus we find that, with the possible exception of standard 'off-the-shelf' supplies, we have to allow two to three clear weeks as a minimum tender period, even for the simplest of tenders. The time of the year, with its effect on holidays, sickness, postal delays and so on may increase this minimum.

**3.5.3** Table 3.1 gives a guide to the average periods needed by engineering contractors to produce a reasonable tender. Within the wide boundaries of the engineering industry there will inevitably be wide divergences. Straightforward constructional works (for which the enquiry includes full drawings and bills of quantities) will usually need considerably shorter tender periods, whilst elaborate 'one-off' machines (probably involving the tenderer in design and even in experimental work before he can arrive at his price) may well take longer than the periods shown. The figures do not allow time for the transmittal of enquiry documents to tenderers, or for return of the tenders themselves. If these are made by post a further two weeks can easily be needed. Figures are for contracts of average complexity in the groups shown.

**3.5.4** Amendments issued subsequent to the enquiry usually involve the tenderer in re-estimating at least (and possibly, in plant contracts, in redesign also) for which an extension of the tender period must be allowed. If the likelihood of substantial amendments can be foreseen prior to the date of the enquiry, allowance should be made from the outset in the enquiry programme but should only be notified to the tenderers as and when the amendments are introduced.

*Table 3.1* Guide to minimum periods to produce tenders

| Nature of contract | Clear weeks to produce tender |
|---|---|
| Supply and erect standard products | |
| Under £35 000 | 2–3 |
| Over £35 000 | 3–4 |
| Supply and erect simple equipment but involving some design by tenderer | 4–6 |
| Supply/erect mechanical/electrical systems, with system design: | |
| Under £500 000 | 6–8 |
| Over £500 000 | 6–10 |
| Construction contract, full bills of quantities supplied with enquiry | |
| Under £1m | 4–6 |
| Over £1m | 6–8 |
| Construction contract with some design responsibility for permanent works | 10–16 |
| Turnkey (multi-discipline), large | |
| Straightforward | 16–20 |
| Complex | 20–26 |

## 3.6   THE ENQUIRY PROGRAMME

**3.6.1**   From the time a decision is taken to go out to tender until the date planned in the project programme for the completion of the resulting contract, there is usually a firm, inelastic period. The progress of the project can just as easily be upset by dilatory launching of the enquiry, or delayed choice of a tender, as by the chosen contractor taking too long to carry out his work. It is easy to squander time at the start, in the mistaken belief that it can all be recovered later on. The enquiry procedure needs a firm programme which the contracts engineer, in conjunction with the planning engineer, must draw up immediately the decision is taken to proceed. It is issued to all Engineer and Employer staffs involved in carrying out the procedure. The programme must be strictly realistic at all stages, due regard being taken of the extent and complexity of each task and the influence thereon of the known state of basic information, for example design drawings, planning permission and the like. Besides key dates for each stage, the programme must identify who is responsible for taking action on time. When unforeseen delay occurs, the contracts engineer must consult with the persons named in the programme to decide how the delay is to be taken up, and must issue a revised programme as necessary. Subsequent stages must not be allowed to 'freewheel' in default of receiving new plans or a new completion date.

The number of detailed steps into which the enquiry programme is divided must clearly depend on its size and complexity. The enquiry programme is intended as a helpful check of the items and the timing to ensure the enquiry is ready by the intended date. It is not expected to be a bureaucratic exercise in form-filling. The example shown at Appendix 5 has been prepared in full detail suitable for the largest and most involved enquiry. It needs to be shortened and condensed for most other enquiries. Though each of the steps will still have to be carried out, they will, in many cases, be shorter and less time-consuming, and hence of less significance in attaining the target date for issue. Some items may be ignored, others can be bracketed together, with a single target date for completion.

**3.6.2** In assessing the period which must be allowed in the programme for each stage of the procedure, the contracts and planning engineers will need to consider, *inter alia*, the following:

- Current loading on each department concerned in preparing the enquiry documents.
- Progress on preparing the final version of the specification.
- Progress on preparation of the enquiry drawings.
- The detail and extent of the bills of quantities foreseen.
- The preparation and/or modification of contract conditions, special conditions, site regulations.
- Decisions affecting the work still awaited (e.g. planning permissions, Employer's decisions or approvals, licences, etc.).
- Progress on preliminary enquiry, tender list, financial status investigations.
- The number of stages which require Employer's approval of proposals and the number of Employer's departments which are involved.
- The task of assembling, typing, copying, collating, packing and delivering documents. Facilities available and their other commitments. Transmittal delays, especially overseas.
- Special actions by tenderers in arriving at their offer, e.g. development work, design, testing or submission of prototypes or models, site investigations.
- Sub-contracting by tenderers. Progress on nominated sub-contracting.
- Additional tasks by tenderers overseas, e.g. investigation of financial and exchange possibilities, shipping and labour availability, translation of documents at all stages, etc.
- Provisional arrangements tenderers may have to make for credit guarantees.
- Extent and duration of negotiations anticipated with preferred tenderers.

- Expected duration of appraisal period and time for assimilation of appraisal report by the Employer.

To meet the realities of life the time for the full procedure should include a contingency of at least 15–20% of the best estimated time, with a minimum addition of 2 weeks. This contingency is kept in reserve by the contracts engineer and not published in the enquiry programme. The less firm the specification and enquiry drawings, the more the contingency needs to be: the less standard the product or straightforward the work, the more certain that specification changes will be introduced with all the delays that they imply.

**3.6.3** If a coldly realistic enquiry programme cannot be made to fit into the time available in the project programme, consideration must be given (in conjunction with the Employer) to possible time savings if one of the other forms of tender were to be adopted (for example, is the contract work a suitable subject for a negotiated tender or for a simple schedule-of-rates tender?). Alternatively, it may be necessary to consider the suitability of the contract works for double-shift or overtime working whereby the contract start date might be put back without delaying the completion date. Whatever the relative disadvantages of these courses may be compared to the disorganization by late completion of an integrated project, at least the enquiry programme should highlight the existence of a problem whilst there is still a possibility of doing something about it.

## 3.7   THE ENQUIRY RECORDS

A few words should be said before leaving the present subject about two pro formae which will prove helpful when dealing with enquiries for the numerous contracts which make up a major project.

### 3.7.1   The Enquiry Log

The Engineer will normally place on one man (usually the contracts engineer) the responsibility for coordination and control of all activities concerned with tendering procedures. This involves the contracts engineer in two sets of actions for which some form of written record is needed:

(a) He must monitor the activities of all other persons and groups concerned in the tender enquiry so that they meet their key dates in the enquiry programme. (He must similarly ensure he keeps his own key dates!)

(b) He becomes the channel through which communications with the

Employer, tenderers and others will flow, and this can involve informal and telephone discussions.

Whoever is so appointed will find it useful, if not essential, to maintain for each enquiry currently in progress a separate Enquiry Log in the form of a loose-leaf diary built up from standard pro-forma sheets of the type shown in Appendix 6. One or more sheets can be added each day and are duly labelled with the date and the serial number of the sheet. The body of the form illustrated is divided into two sections intended to meet the two activities (a) and (b) above.

The first section is intended for reminders by the contracts engineer of key actions due to take place that day, either by himself or by others concerned in carrying out the enquiry programme for which he is responsible.

The second section is for recording as they take place any significant events, conversations, discussions, verbal decisions and the like which might otherwise not appear on the project files as letters or other written documents. Entries recorded in the log at the time of the event can also be most useful when drafting or checking written confirmation of such verbal matters later.

We shall see in Chapter 10 that the contracts engineer is responsible for maintaining throughout any contract the *Master Contract Record*, a document which is relied on to produce complete and reliable data in the event of any disagreement or litigation between the Employer and the contractor or with a third party. The contracts engineer should rely on his Enquiry Log to bring into the Master Contract Record events, decisions and the like which might otherwise go unrecorded.

### 3.7.2 Record of enquiries issued

There is a requirement for a single-sheet record of the addresses to whom an enquiry has actually been sent. From the provisional tender list onwards, the roll of tenderers can change frequently and repeatedly, with the details becoming submerged in inflated tender files. The record of enquiries issued serves as a consolidated reference sheet. A suggested pro forma is shown at Appendix 7 and is self-explanatory.

### 3.8 FLOW OF DOCUMENTATION

The sequence of actions during the enquiry procedure can be well summarized and illustrated by the flow charts for documents between the persons involved. They are shown diagramatically in the two accompanying charts which cover the procedure during two basic periods: Table 3.2

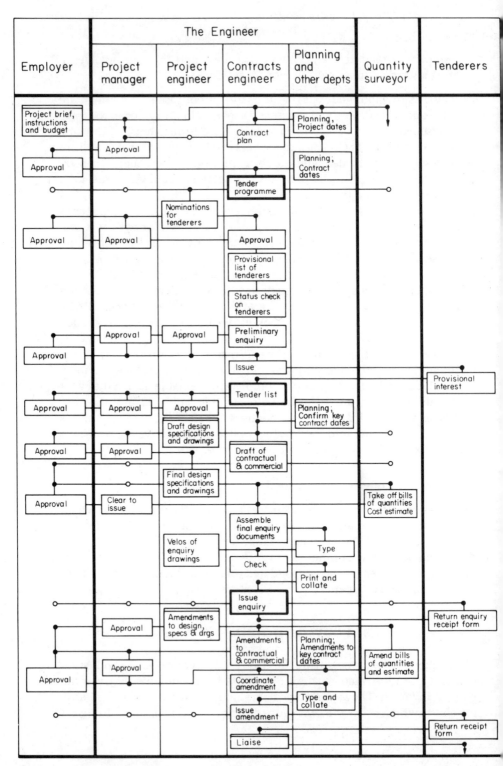

*Table 3.2* Flow of documentation – Enquiry period

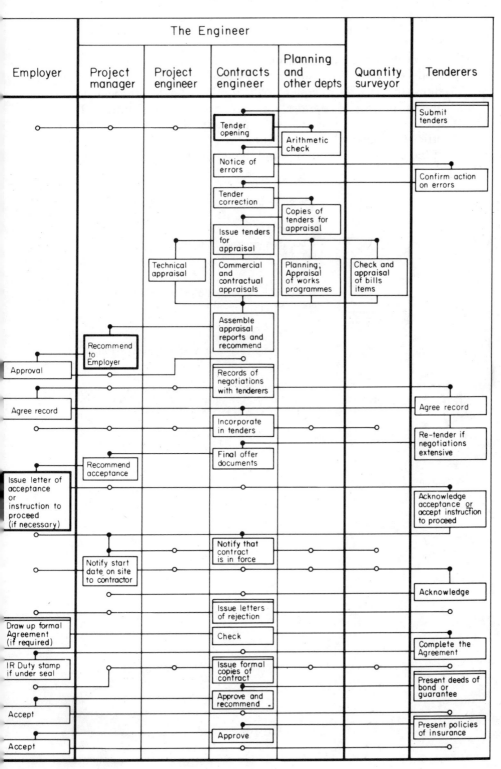

Table 3.3 Flow of documentation – Negotiation period

shows the enquiry period up to the date fixed for the submission of tenders (which is usually the tender-opening date also), and Table 3.3 shows the period subsequent to the receipt of tenders through negotiation and the process of appraisal up to the placing of the contract. The charts also serve as a ready-reference to many of the duties of the various officials concerned with operating the procedure. Their titles are shown at the heads of the columns and the following symbols are used:

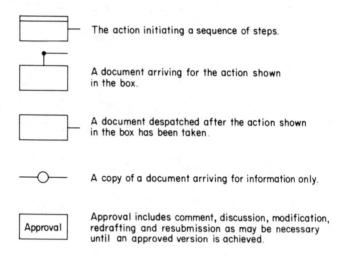

The action initiating a sequence of steps.

A document arriving for the action shown in the box.

A document despatched after the action shown in the box has been taken.

A copy of a document arriving for information only.

Approval includes comment, discussion, modification, redrafting and resubmission as may be necessary until an approved version is achieved.

# 4  The enquiry documents

## 4.1  THE DOCUMENTS FOR THE ENQUIRY

An enquiry inviting tenders will include the documents which are described in the remainder of this chapter:

- Covering letter (the invitation to tender).
- Instructions to tenderers.
- The enquiry documents.
- Completed labels for the return of tenders.

The enquiry documents vary in detail and scope depending on the subject, nature and size of the contract works but usually include many, or all, of the following:

- Declaration of bona fide tender.
- Formal form of tender.
- Data sheets and pro formae for additional information the tenderer is required to provide.
- Contract programme – key dates.
- Conditions of contract (with any special conditions introduced by the Employer/Engineer).
- General specification of broad requirements of the works.
- Site data and facilities.
- Technical specification of the works.
- Drawing list and relevant drawings.
- Schedule of nominated sub-contractors.
- Site regulations.
- Preamble to the bills of quantities.
- Bills of quantities (or schedule of rates) including any provisional sums and prime cost items.

## 4.2 RESPONSIBLITY FOR THE PRODUCTION AND ISSUE OF THE ENQUIRY

**4.2.1** The Engineer's responsibility for the coordination of the work on the enquiry and the assembly and timely issue of the documents to tenderers is usually placed by the project manager on his contracts engineer. The contracts engineer will, himself, be directly concerned in the drafting and preparation of certain of the documents (see Section 4.2.2) and he must monitor the production of the remainder to ensure they are available in the final form in time for reproduction and issue. The contracts engineer's duties will therefore include:

- Preparation of the list of tenderers.
- The enquiry programme.
- Monitoring and chasing action to ensure preparations are running to time.
- Preparation of those documents for which he himself is responsible.
- Receipt of master copies of final enquiry documents produced by others.
- Co-ordinating matters of interfacing and overlap in the several documents to avoid ambiguities or omissions.
- Submission of set of final enquiry documents to Employer for approval of issue.
- Reproduction, collation and issue of the enquiry.

The implementation of these responsibilities is illustrated by Table 3.2 and many of the items above are referred to in more detail elsewhere in this book.

### 4.2.2 The preparation of the separate documents and data

The distribution of responsibility among members of the Engineer's team for production of the individual parts of the enquiry documentation is usually divided up as shown below. This division does not imply that the engineer named produces the document on his own. In many cases the contents are the result of discussions between (or include contributions by) other members of the team and their opposite numbers on the Employer's staff. The liaison with the Employer will follow the principles established at the outset of the project, such as we have considered in Section 2.1.4, and in particular paragraph 2. The contracts engineer, in his role of coordinator, deals with matters affecting the enquiry as a whole.

The sub-division of responsibility for production of the individual parts of the enquiry will normally be:

*Project engineer:* nominations for inclusion on list of tenderers; general specification (broad description of the works); technical specifications from all disciplines concerned; enquiry drawings (all disciplines); site data; nominated sub-contracts (with contracts engineer).

*Contracts engineer:* letter of invitation to tender; instructions to tenderers; form of tender; pro formae for additional data (in conjunction with project engineer); conditions of contract (including special conditions); commercial terms, terms of payment, rates of liquidated damages, etc.; site regulations (in liaison with Employer); nominated sub-contracts (with project engineer); arrangements for issue of enquiry and receipt of tenders.

*Planning engineer:* contract programme (including start date and date for completion); key dates imposed on progress of the works by other parts of the project; part completion dates for co-ordination with other contracts, and their relative importance to the project.

*Quantity surveyor:* preamble to bills of quantities; bills of quantities (or schedule of rates); specification of method of measurement; estimates for provisional sums and prime-cost items; pre-contract estimate for the works.

In the event that the Engineer's organization is unable to call on adequate quantity surveying effort, the Engineer will retain a firm of quantity surveyors to carry out the work. The principles of liaison already described in Section 2.1.4 above will apply to their link with the Engineer's staff, namely with the project engineer on technical matters affecting the bills or the pricing of items, with the contracts engineer on matters of commercial or contractual significance and with the project manager on matters affecting policy or the Employer's intentions.

## 4.3 DISCUSSIONS WITH TENDERERS REGARDING THE ENQUIRY DOCUMENTS

A principle of competitive tendering is that all tenderers have the same opportunity to make offers on the same basis for the same work; their offers are then directly comparable. Discussion with individual tenderers during the tender period upsets the balance of equal opportunity and must be avoided wherever possible by making the enquiry sufficiently comprehensive and precise to stand on its own.

What this requirement implies in practice varies considerably with the nature and extent of the contract. In the purchase of specialist electrical or mechanical equipment, for example, the enquiry will usually be based on a functional specification leaving the tenderer maximum freedom (as the specialist in the field) to submit his proposals. The 'competition' element is not only on price but on technical excellence as well. On the other hand in building or civil engineering contracts the specification is usually in full detail, supported by drawings and complete bills of quantities. The competition here is essentially a matter of the constructional methods and organization proposed by each tenderer, which reflect in their standard of work and time taken (i.e. price). The contractor is given no scope on the design of the permanent works, but may exercise his ingenuity on the temporary works he proposes to use.

Architectural drawings are customarily in more detail and the bills of quantities taken off them can be more precise, with the result that they are, together, often sufficient fully to specify the works, making a separate specification document superfluous. With enquiries for building and civil engineering contracts therefore there should be little need for queries from tenderers on the documentation except for an interpretation of the occasional misprint or ambiguity which escaped detection. Queries are likely to be more concerned with the implementation of the constructional methods the tenderers propose to adopt and their acceptability.

With enquiries for plant, however, each tenderer may have queries as to whether certain embodiments or points of design would be acceptable, whether certain refinements in plant detail would be welcomed and the extra cost thereof accepted, and innumerable other aspects to ensure their technical proposals are in accordance with the Employer/Engineer's views. It is so easy for a tenderer to price himself out of the contract by putting forward refinements which the Employer/Engineer does not consider worth spending money on. In such enquiries it becomes a problem during negotiations for the Engineer to treat all tenderers the same, when their solutions can be so diverse. Considerable care has to be taken to maintain confidentiality regarding individual solutions exposed by tenderers during the tender period. The Engineer must ensure that any *relaxation of or modification to* the functional specification which is permitted to one tenderer during the course of such discussions is promptly notified and permitted to all the others whether or not they have requested it.

*Embellishments* to a functional specification by one tenderer are to be regarded as his confidential solution, not to be divulged to the other tenderers. Such embellishments will usually cost extra and the tenderer's competitive position will depend on how the advantage of technical excellence will stand in respect to a disadvantage in price. The criterion is still value for money.

## 4.4  SCOPE OF THE ENQUIRY DOCUMENTS

In Sections 4.5–4.18 we shall examine the scope of each item making up a typical set of enquiry documents for a large contract. Because every enquiry has its own special features, the feasibility of making appropriate modifications to the descriptions given must be borne in mind. The check lists we have included have been made fairly comprehensive so that, it is hoped, such 'modifications' will in most cases amount to 'omissions'.

## 4.5  THE LETTER OF INVITATION TO TENDER

This is the covering letter to the tenderer, accompanying the enquiry documents. It needs to be short, formal and for the most part stereotyped. If the addressee has previously responded favourably to a preliminary enquiry (see Section 3.2) this should be referred to with dates and references. The letter should also include (except in so far as such matters are covered in the instructions to tenderers):

- Name of Employer and scope and location of project.
- Tender number and title (i.e. what specifically the enquiry is for).
- Name of Engineer's or Employer's representative to whom queries should be addressed and where he is to be contacted.
- Date, time and place for return of tender (or refer to instructions to tenderers).
- A receipt form for the enquiry documents to be signed and returned. This should schedule all the documents as a check list: if it does not, the list should be quoted in the letter itself. The letter must be signed by the project manager or by the contracts engineer on his behalf.

A typical invitation to tender is shown at Appendix 8.

## 4.6  INSTRUCTIONS TO TENDERERS

**4.6.1**  As the name implies this document tells the tenderer what he must and must not do when compiling and putting forward his offer. It is also used to draw his especial attention to features of the enquiry which are unusual and might be overlooked in the general mass of enquiry documentation, but which could have a considerable bearing on his offer. The title itself implies that the document will have served its purpose once the tender has been submitted: it would not be included in the eventual contract documents. It is not usual, therefore, to consider it a formal tender document. This aspect must be borne in mind when deciding if any particular item should be included in the 'Instructions to Tenderers' or is more properly dealt with elsewhere.

**4.6.2**  Some or all of the following points might need to be dealt with in the 'Instructions to Tenderers':

(a)  As instructions which must be complied with:

- Which enquiry documents have to be filled in by the tenderer, and which signed.
- Whose signature to the tender offer will be acceptable (e.g. a Director or Purchasing Manager).
- Any rules to be observed in pricing (e.g. prices to be those ruling at a stated base date, divisible tenders, treatment of Value Added Tax, charges, duties, taxes, etc., to be included).
- Validity period required.
- Any documents other than enquiry documents which the tenderer must include with his offer (e.g. 'fair wages' certificate, Finance Act 1971 certificate, technical descriptions or drawings, proposed methods of construction, proposed progress chart).
- Which documents have to be returned with the offer, and how many copies of each.
- The date and hour by which tenders must be in the hands of the Engineer.
- Instructions for packing labelling and addressing the tender (including a supply of standard labels – see Section 5.2.3).
- Named person to whom queries arising from the enquiry documents must be addressed and where he may be contacted.
- Procedure to be adopted if tenderer wishes to inspect the site of the works.
- Procedure for inspecting any special reports or plans not circulated with the enquiry (e.g. site investigation reports).
- Circumstances under which alternative or non-compliant offers may be submitted.
- No alterations to be made to enquiry documents: any points of non-compliance to be scheduled separately.

- Confidentiality of enquiry documents.
- In the event no tender is submitted, all copies of the enquiry documents are to be returned to the Engineer.

(b) Drawing attention to unusual features which are dealt with in the enquiry documents proper, for example:

- Currency or currencies in which tenders must be priced. Quoting of exchange rates assumed.
- Situation regarding price adjustment and the relevant formula to be used.
- Whether quantities are fixed or subject to remeasurement.
- Restrictions on freedom of operation on site or access to site.
- Potential interference from other contractors who will be working on or adjacent to the site at the same time.
- Situation regarding provisional or prime cost items.
- Nominated sub-contracts (known or envisaged) in a main contract (see (d) below).
- Applicable law, if not that of England.
- Bonds or guarantees required from the successful tenderer.
- Insurance cover to be taken out by the successful tenderer: proof to be produced.
- Submission of data by tenderer as called for in the enquiry will not imply approval by the Employer or limit the contractor's obligations.
- Any other unusual features of the particular enquiry due to its special conditions of contract.

(c) Information on procedures the Engineer will adopt in connection with tenders, for example:

- Procedure for the issue of amendments to the enquiry if these are necessary.
- Procedure for dealing with requests for extension of time by tenderers.
- Procedure for dealing with arithmetic errors found in tenders after submission.
- Rules relating to disqualification/rejection of tenders (late arrivals, altered figures, incomplete, etc.).
- Employer not obliged to accept the lowest, or any tender.

(d) If the enquiry in question is in respect of a nominated sub-contract, certain special points need to be be made linking it with the main contract, for example:

- Details of the intended main contract and contractor (if known).

- Where copies of the intended main contract can be examined by tenderers.
- Need for compliance with the conditions of contract of the main contract: what the standard conditions are, details of any amendments thereto, any special conditions, site regulations, etc. (relevant matters should not only be referred to, but reproduced as part of the sub-contract enquiry documentation).
- The successful tenderer will be required to enter into negotiations with the main contractor on all matters except price, and to conclude a nominated sub-contract with him.
- The main contractor may require a sub-contract performance bond as expressed in the main contract.
- Whether the Employer requires a collateral Agreement with sub-contractor regarding design responsibility and the avoidance of delays to the sub-contract dates.

**4.6.3**  The Health and Safety at Work Act 1974 places added responsibilities on the Employer (and the Engineer), the contractor and the contractor's workmen in ensuring adequate safety measures are carried out on site. Contractors are required by Section 2(3) of the Act to have a set of safety rules in respect of their own operations on site and it is customary for copies of this to be handed to each workman employed on site to ensure he knows what is being provided for him and what is required from him. It is recommended these safety rules are examined before a contract is let to ensure they are both adequate themselves and in conformity with the Employer's site regulations. Tenderers should be instructed to forward a copy with their tenders, without which a tender may not be considered.

## 4.7   DECLARATION OF BONA FIDE COMPETITIVE TENDER

This declaration is usually required to be included in enquiries by or on behalf of public authorities but more recently has also been adopted to a growing extent by other Employers. The wording of the declaration has become virtually standard in the form used by many public authorities, and this is shown at Appendix 9. If done deliberately or negligently, the making of a false declaration would be a fraudulent misrepresentation, which under English Law would entitle the Employer to rescind any contract entered into with the declarer, and to claim such damages as he had incurred as a result.

A refusal by a tenderer to sign the declaration might suggest the existence of a collusive arrangement and could be of significance in proceedings under the Restrictive Trade Practices Act (1968).

## 4.8  THE FORMAL FORM OF TENDER

**4.8.1**  A tender as submitted by the tenderer must include a formal offer to do certain things for a stated price: this offer, on acceptance by the Employer, constitutes the contract. Offers worded by tenderers themselves are frequently incomplete, imprecise and legally wanting, and differ in content one from another among the competitors for a given contract. It is important therefore that with all enquiries (with the possible exception of minor 'supply only' contracts) a uniform, legally adequate pro forma offer be issued with the enquiry documents and all tenderers be required to complete and sign it. This is the *Form of Tender* and its format differs in some respects between three classes of tender:

- Where the works are firmly specified (even if only by function) and a firm tender price is meaningful.
- Where the works cannot be firmly specified, as for example development contracts, cost-plus contracts or construction contracts subject to remeasurement. Here the price to be paid is determined in accordance with the terms of the contract probably based on established schedules of rates. A lump-sum tender price has no contractual significance.
- Where the tenderer is intended to enter a nominated sub-contract with a main contractor (yet to be appointed by the Employer) at the same prime cost as quoted to the Employer in the form of tender.

A typical form of tender for each class is given in Appendix 10. Some standard forms of conditions of contract also incorporate a generalized form of tender but normally these are not specific enough for most enquiries without a number of additions and amendments.

The third class of form of tender is a very good illustration of the weaknesses of the nominated sub-contractor concept. The offer is made to the Employer, who (in the absence at this time of a main contractor) can only make it binding on the tenderer by himself accepting it and entering a formal contract direct with the 'sub-contractor'. He therefore introduces an agreement by the tenderer that he may later at his discretion, transfer this contract to his main contractor. In doing so he cannot re-open negotiations and therefore he seeks to make the terms, conditions, special conditions and site regulations applicable to the enquiry such as will (he hopes) prove acceptable to the main contractor when invited to take over the nominated sub-contract from the Employer. Usually the aim will be a back-to-back reflection of the terms and conditions he hopes to agree with the main contractor – but as these negotiations may not at this stage even have started, it is largely a matter of guesswork.

Such back-to-back terms may be wholly inappropriate to the nature of the sub-contract works and the usage of the industry involved, for example an electronically-controlled traffic signal as part of a major road construction project governed by ICE Standard Conditions of Contract (5th edition). The tenderer may not be prepared to quote against such terms and conditions.

Finally, as a backstop, the Employer invites the tenderer to bind himself by the form of his offer *to enter negotiations* with a future main contractor (whoever he may be) for a nominated sub-contract at his tender price and the terms, conditions and regulations set out in the enquiry. He cannot be invited *to agree to enter such a contract* as such an 'agreement to agree' is legally invalid: the Employer can do no better than ensure negotiations will start even if they rapidly prove to be abortive.

In Section 8.2 we deal in greater detail with the setting up of nominated sub-contracts, but the form of tender brings into clear focus some of the contractual problems which underlie this well-established method of working.

**4.8.2**   The form of tender, being the legal offer, clearly needs drawing up with some care, to ensure all matters necessary to define the tender are included, and this is usually the responsibility of the contracts engineer.

- The *statement of offer* (paragraph 1) must, by direct reference, incorporate all the documents to which the offer is subject. When the tender is accepted they will become contract documents.
- The *'consideration'* (normally the price payable) must be quoted in figures or as a specified basis for calculation.
- The *contract programme* (see Section 4.10.5 below) must be introduced.
- The *validity period* during which the tender offer is kept open for acceptance must be defined.
- The tenderer must undertake to complete a formal Agreement subsequent to the contract being awarded (if this is required) and to recognize a contract formed by acceptance of the tender to be fully valid until the formal document is executed.
- There should be included an undertaking to enter into any bonds or guarantees called for by the contract documents. The latter may be legally sufficient, but the undertaking confirms its importance.
- If the Employer wishes to retain an option to let the contract works in sections to more than one contractor, the tenderer must state his

willingness to accept part only of the works at the Employer's discretion. The tenderer must quote for each section of the works separately and to say the price reduction he will allow on the total of such sections, if the whole contract works are ordered from him.

- It is usual to include recognition by the tenderer that there is no obligation implied by the enquiry for the Employer to enter into a contract with anyone.

**4.8.3** Random alterations by tenderers to the form of tender (or indeed to any other enquiry document) cannot be permitted: it would mean a thorough search through every tender from end to end to reveal them. If a tenderer is unable to make an offer which is fully compliant with the enquiry as issued then he must introduce, preferably as an appendix to the form of tender, a schedule of all points of non-compliance, technical, commercial and contractual to which his offer is subject (See Appendix 18). The schedule must show clearly for each item:

- The exact points and extent of the enquiry documents (as issued) which are not complied with, and the reasons.
- What is offered in substitution.
- A statement of the operative effect of the substitution (where this is not immediately and fully obvious) on the specification, performance and price.
- Whether the substitution is of the essence of the offer or is open to negotiation.

## 4.9   PRO FORMAE FOR ADDITIONAL INFORMATION

The Engineer frequently wishes to ensure that tenderers submit additional information about their offers, covering specific aspects. To ensure completeness and uniformity between tenderers, appropriate data sheets or pro formae are issued with the enquiry. The instructions to tenderers (or elsewhere in the enquiry documents) require all tenderers to complete and authenticate the forms as an essential part of their offers. Authentication can be achieved in one of two ways.

- Each form, standing on its own, is individually signed by the tenderer. This has two disadvantages, first, the not inconsiderable task of signing all the forms, and secondly that they are not automatically incorporated into the offer unless the form of tender is elaborately worded to ensure this. Thus they may later fail to achieve the force of a contract document.

- Each form is designated as a serially numbered appendix or 'part' of the form of tender and is incorporated by number in the 'offer' of the form of tender itself. In this way each 'part' or appendix becomes an intrinsic part of the offer and eventually of the contract documents. The signing of the form of tender itself avoids the need for signing every appendix or 'part' separately.

It is still good practice for the tenderer to identify each 'part' pro forma as being the relevant one by initialling it at the same time as the form of tender is signed. This is not normally made a requirement, at any rate for home contracts. The more commonly used pro formae include the following, some of which are illustrated by typical examples in the appendices hereto.

| *Pro forma* | *Appendix* |
|---|---|
| • Breakdown of lump-sum tender price (e.g. for appraisal, part-awards or CPA calculations). | |
| • Schedule of proposed sub-contractors for approval. | 16 |
| • Data relating to authorized overtime and daywork. | 11 |
| • Contractor's electrical power requirements on site. | 12 |
| • Schedule of prices for use with variations to plant requirements. | 14 |
| • Schedule of plant-hire rates (for day-work). | 11 |
| • Schedule of plant to be brought on site by contractors. | |
| • Projected build-up of labour on site. | |
| • Schedule of all items not complying with enquiry. | 18 |
| • Projected road traffic – vehicles (by class) entering or leaving site daily. | |
| • Schedule of recommended holding of consumable spares (priced) (for plant being tendered for). | 13 |
| • Schedule of recommended holding of maintenance spares for 2 years running (for plant being tendered for). | 13 |
| • Accommodation in site camps required for employees of different classes (overseas contracts). | |
| • Provisional programme for execution of the contract works. | 15 |
| • Contractual times for completion. Schedule of rates of liquidated damages for delay in keeping them. | 17 |

## 4.10  DATES OF SIGNIFICANT CONTRACTUAL IMPORTANCE

Each contract features a number of dates which, for various reasons concerned with the execution of the contract, have a legal significance: the terms of the contract must be so expressed that they can be identified and established precisely. As far as possible, dates should be specified at the time of the enquiry, but where this is not possible they should be established in terms of elapsed periods from a stated base date, such as the *Contract Date*. As soon as a provisional date becomes firm (for whatever reason) or a firm date is extended by the Engineer under a provision of the contract conditions the new date should be put on record by the Engineer, and both parties notified in writing.

### 4.10.1  The contract date

This is the date when a valid contract legally comes into existence. Normally it will be the date on which an unconditional acceptance is given to an offer (or counter-offer). In most engineering contracts, the acceptance can be verbal or written. If it is later *confirmed* in writing or by the execution of a formal Agreement, the contract date is still that of the original acceptance.

### 4.10.2  The effective contract date

Although a binding contract may have come into existence legally, the parties may previously have agreed in its terms and conditions that its execution by both parties shall not begin until some specified later date, the effective contract date. One reason for this might be simply that the site is not available at the contract date for work to start (see Section 4.10.3), but frequently the delay is imposed by some contingent action, essential to the operation of the contract, which could not be put in hand until a binding contract actually existed. Examples might be the granting of permits or licences: a more complete list is given in Section 9.1.1.

The effective contract date must be specified, either as an actual date or (more usually) as the date on which the last delaying contingent action is completed. It is then used (instead of the contract date) as the base date for the purpose of calculating contractual periods to delivery, completion and the like, and the contractor's liability for damages for delay.

In parenthesis it may be noted that the conditions may further provide that the contract shall become void if any of the named contingent actions proves impossible of performance within a stated period after the contract date, e.g. 6 months.

### 4.10.3  The start date

In contracts in which work on site must take place from the outset, the start of operations may be delayed by the non-availability of the site itself, or of access to it, or by some earlier contract (on which the present one depends) not having made adequate progress. In such cases the contract may need to specify a *Start Date* other than the contract date, and this should be stated in the enquiry whenever it can be foreseen by the Engineer's planning department.

In a complex contract, separate start dates may be specified for different sections of the works to enable them to tie in with concurrent operations by other contractors.

Contracts for civil engineering or building work are the ones usually affected. Indeed it is customary in such contracts to identify a start date some short period (7–14 days) after the contract date even if there are no external limitations, to enable the contractor to get himself and his equipment, labour, and supplies mobilized for a good start on site.

Contracts for the supply of plant seldom, if ever, call for a start date as such, the contractor starting work in his own factory at such time as he chooses. They may, however, still have written into them limiting dates before which erection or installation of the plant on site cannot begin.

A start date differs fundamentally from an effective contract date. Prior to the latter, direct action on the contract by both parties is in abeyance; prior to a start date it is only work by the contractor on site which is postponed: everything else can go ahead. In construction contracts which are essentially all site work, the whole of the contract is usually timed from the start date and not from the date of contract.

### 4.10.4 The completion date

The completion date usually has considerable contractual importance and therefore needs to be closely defined in the contract so that there is no room for doubt when it has been reached, or looked at another way, the exact date at which the contractor becomes in delay. In the various standard forms of conditions of contract the definition of completion varies slightly, but they mostly signify that the contract works have to be finished to specification, tested and/or demonstrated to the Engineer that they are so. The Employer then takes over the works from the contractor whose responsibility for making good defects under warranty begins. A substantial part of the contract price is usually payable at completion. If the contractor fails to complete by the completion date (or such extension of it as the Engineer may have approved and authorized under the terms of the contract) he is usually in breach of contract and is then liable to pay the Employer damages incurred as a result. Completion dates should be quoted in the enquiry documents, preferably as actual dates, though it will frequently not be possible to fix them so accurately at the enquiry stage. If dates are stated they must be associated with a start date by which it is to be assumed the contract will have been let, otherwise dates are replaced by periods (in elapsed weeks) between start and completion (or both referred to a recognizable zero week such as that containing the date of signing the contract). It may be necessary to define how the intervening periods are calculated (see for example Appendix 17) so that there is no ambiguity as to the starting and finishing dates.

### 4.10.5   The Contract Programme

The Contract Programme is a contract document recording dates which the contractor and Employer have bound themselves to meet. It will thus record the contract date, the start date and completion date both for the contract as a whole and for any sections of the works which have to be ready by definite times in order to work in with the rest of the project. It may, for similar reasons, embrace such dates as delivery to site or of arrival by sea at local harbours.

The Contract Programme is not to be confused with any plans, programmes or charts (such as the tenderer's provisional programme, Appendix 15) which the contractor may have to produce for the Engineer to show his plans for carrying out the works or, later on, to aid supervision and checking of day-to-day progress. Although these are useful control mechanisms for both parties, the contractor is under no contractual obligation to adhere to them, provided he meets his contractual dates in the contract programme itself.

The Contract Programme should be one of the enquiry documents, a typical form being shown at Appendix 17. Some of the standard forms of conditions of contract also have suggested forms arranged to conform with the appropriate clauses of the conditions.

Such forms allow for the division of the works into sections, each with its own contractual dates. Some may include the agreed liquidated damages payable on failure to meet the dates concerned, each item being allotted a rate of damages estimated to represent the true damages the Employer might be expected to suffer by a delay in meeting the contract date by the contractor. The amounts can clearly be very different for the various sections of the works, depending on the contingent project works which are affected. Rates of liquidated damages can, by agreement between the parties, be less than, but must never exceed, the justifiable estimates referred to: if they do, they cease to be a recovery of damages and become penalties (i.e. punishments) and as such are not admissible under English law.

## 4.11   THE CONDITIONS OF CONTRACT (INCLUDING SPECIAL CONDITIONS AND CONTRACT PRICE ADJUSTMENT CLAUSES)

**4.11.1**   The Conditions of Contract are the rules by which the execution of the contract is to be governed. They set out the responsibilities, rights and liabilities of the two parties, and the actions to be taken by the parties if and when certain eventualities should arise. They also define the authority and duties of the Engineer in controlling the contract and the actions of the contractor. A full appreciation of the scope and interpretation of conditions of contracts is a matter for a specialist contracts engineer or a lawyer and is not appropriately dealt with here. It is, however, as well to explain that the terms are considered legally to be of two degrees of importance, depending on the actual wording, the subject and the circumstances of any given contract:

*Conditions:* terms expressing matters basic to the contract. A failure to perform their requirements implies breach of an essential obligation, leaving the victim with something differing in an important way from that which he contracted for.

*Warranties:* terms dealing with matters not of the essence of the contract, being subsidiary to the main purposes for which the parties contracted.

If a party fails to fulfil any term of the contract he is said to be 'in breach' thereof, and the other party may claim any damages he suffers as a result. If the term in breach is regarded as a 'condition' the victim may also treat the contract as at an end (a thing he cannot do in the case of a breach of warranty). He is not forced to bring the contract to an end, so that unless he does so at once, and himself also ceases to act on the contract, he will be deemed not to have applied the stoppage, but to have affirmed the contract. He may still, however, claim damages.

It is important to distinguish between conditions of contract as defined above and site regulations (see Section 4.17). The latter are administrative rules for the use of the site, which both parties are expected to observe for reasons of security, orderliness and general convenience, and to preserve the facilities and amenities of the site itself.

### 4.11.2   The conditions of contract proposed in the enquiry

Although either party may propose the conditions which are to apply to a contract, the Engineer, when sending out an enquiry, will normally state the ones which he intends shall govern the resulting contract, and the tenderer is expected to comply with them. They will usually comprise:

- A standard form of general conditions of contract appropriate to the nature of the works involved.
- A series of amendments to the foregoing, in order to adjust them to the circumstances of the actual contract concerned (see Section 4.11.5).
- A number of special conditions of contract which deal with matters peculiar to the contract and not dealt with by the standard form (see Section 4.11.6).

If a tenderer objects to any or all of the clauses he must record the fact in his tender (see Section 4.8.3) and specify what he wishes to put in their place. He may propose further amendments, additions or deletions, or more extremely, he may make his offer contingent on a quite different set of conditions, either his own or another standard form more appropriate to his work. Such a drastic change is more likely to occur with tenders for the supply of plant or highly specialized equipment (e.g. electronic equipment, steam turbines, etc.) for which particular sets of contract conditions exist, which such a supplier would normally use.

If the offer proves to be otherwise interesting, and the tenderer is selected as one of the preferred tenderers, it will then be a matter for negotiation to arrive at a compromise solution satisfactory to both parties (and to the Engineer!).

### 4.11.3   The standard forms of conditions of contract

Of the numerous standard forms of conditions of contract, those in most common use in engineering contracts in the UK have been drawn up and published, individually or jointly, as model conditions by:

- The engineering institutions.
- Trade associations.
- Professional associations and institutions.
- Government departments.
- Nationalized industrial groups.
- Large industrial firms.
- International federations.
- UN Economic Commission for Europe.

Without wishing to make invidious distinctions, the most common of these in the engineering field are:

Civil engineering and construction contracts
    Institution of Civil Engineers General Conditions
    Joint Contracts Tribunal Standard Form of Building Contract ('JCT' or
    'RIBA') (six versions)

Plant, machinery and equipment (supply with/without erection)
    IMechE/IEE Model Conditions (six major versions)
    Institution of Chemical Engineers (Model Form for Process Plants)
    The British Electrical and Allied Manufacturers Association (seven
    major versions – use not confined to electrical equipment contracts)

International use
    Federation Internationale des Ingénieurs-Conseils (FIDIC) (two versions
    for civil engineering and for machinery based on ICE (5th) and
    IMechE/IEE (B.3))
    UN Economic Commission for Europe (principally types 188 and 574 –
    or 188A and 574A with erection)

In the UK each in its own sphere, the following will be met with
    Government Contracts GC/Stores/1 and GC/Works/1
    British Electricity/BEAMA Extended Form 'A' or 'B'
    CEGB Conditions of Contract
    National Coal Board
    British Steel Corporation
    etc.

No attempt has been made to quote references to current editions as these
are constantly under review, and might only become out of date and
confuse the reader.

Each set of conditions is, of course, written primarily to serve its own
branch of engineering, so that problems readily arise when an enquiry
embraces more than one engineering discipline. We have already referred
to this in the Introduction, and we shall meet it elsewhere in the text. The
trouble is especially marked when the disciplines use basically differing
methods of contract operation, for example involving a construction
project with machinery supply and installation. In no circumstances can
two standard forms of conditions be used together on the same contract:
their provisions will inevitably conflict with one another and result at
best in ambiguities and some chaos, and at worst in the contract being
void on grounds of uncertainty. The situation can only be met by choosing
the more appropriate conditions (i.e. those designed to fit the major parts
of the works) and, by skilful amendments and additions, adjust them to
embrace the needs of the lesser discipline as well. This is a task for a
practised contract draughtsman.

We might, at this stage, also draw attention to the difficulties which arise when the Employer, the tenderer, or both, are located in a foreign country using a different legal code, on which they insist on basing their contract conditions. Many of our standard conditions, compiled around English law, will be unacceptable, as (in many cases) will be the commoner international ones. It may become necessary either to accept an unknown standard from the country concerned, or to draw up *ad hoc* conditions. In either case, there is absolutely no alternative to obtaining assistance from a native lawyer practising in the country concerned.

The special case of conditions of contract to apply to nominated sub-contracts is dealt with in Section 8.2.

### 4.11.4   Completion of standard forms of conditions of contract

Most of the standard forms of conditions of contract contain one or two clauses which require completion by the Employer/Engineer (in the light of the project circumstances) before the conditions are issued with an enquiry.

Unless such decisions are made and incorporated, the clauses are usually meaningless. The more common points (though by no means the only ones) are:

- Schedule of recognized sections of the works (as defined in the conditions of contract).
- The rates of liquidated damages payable on delay and any maximum limits to the total sums so payable.
- An agreed arbitrator or the method by which one is to be chosen if and when required.
- The period of maintenance ('warranty period').
- The minimum amount for which claims for interim certificates or payments will be accepted. Percentage of contract values of the works to be included on interim certificates.
- The requirements and value of any performance bond.
- Minimum indemnity to be covered by the contractor's public liability insurance.
- Proposed nominated sub-contractors.
- Percentage for adjustment of PC sums.
- Period of *force majeure* delay before contract can be determined.
- Definition of *force majeure* to apply.
- Whether contract price adjustment is to apply or not and the formula to be used. The base date to be used when quoting.
- The contract programme – start dates and completion dates.

In addition it is usually necessary to insert the name and address of the Employer and of the person designated as the Engineer for the contract. It must be remembered that the Engineer has no authority or standing in relation to the contract except in so far as it is given to him under the conditions of contract (either in the standard form or suitably modified *ad hoc*). He is usually empowered thereby to nominate representatives, assistants, clerks of works, as appropriate but this authority must be checked to ensure it is adequate for the running of the contract (for example, if a quantity surveyor is needed in a contract governed by the IEE/IMechE Model Form 'A' Conditions, a special clause covering his appointment and responsibilities would be required).

In the special case where an enquiry is being issued for a nominated sub-contract (in which the sub-contractor will be expected to accept, in general, the same terms and conditions as will exist between the Employer and the main contractor – his customer) it will be necessary not only to tell the tenderers for the sub-contract what standard form of conditions will be used in the main contract, but also to give them the completion data described above, together with any amendments introduced and special conditions added. Certain other features must also be included and these have been indicated already in Sections 4.6.2(d) and 4.8.1.

### 4.11.5   The amendment of standard forms of conditions of contract

The use of a standard form of conditions must never be regarded as a substitute for thought. Not only may the peculiarities of a contract make some clauses unsuitable or impracticable as they stand, but in most standard forms some items are purposely left indefinite in detail so that the commercial views of the parties may be readily introduced (see Section 4.11.4). These points must be recognized by the Engineer's staff and suitably amended or completed by the contracts engineer, in conjunction with the Employer, before an enquiry is issued to tenderers.

Standard forms of conditions constitute a co-ordinated professional document. The clauses are mutually supporting and closely interlinked, with the result that a simple change to one clause may necessitate changes to others. Otherwise ambiguities, inconsistencies and outright conflict can arise. Take a simple example:

The standard form of conditions says 'delivery of the machinery will be made by the contractor to site'. Because he knows the site access road will not be ready in time, the Engineer alters this to read 'delivery of the machinery will be made ex contractors works'. This simple change may well call for adjustment in the clauses dealing with the following:

- Definitions (i.e. of 'delivery').
- Authority for contractor to deliver.
- Responsibility for provision of transport.
- Transfer of risk in the machinery.
- Transfer of ownership in the machinery.
- Marking of the machinery.
- Recovery in the event of bankruptcy.
- Responsibility for unloading and placing in position on site.
- Insurance.
- Instructions to contractor to install.
- Warranty period.
- Testing before despatch.
- Acceptance tests after erection on site.
- Terms of payment.
- Storage and making good deterioration.

It follows that the introduction of amendments into standard forms of conditions is a matter for those with experience in such work and it should be left to them: a little learning is, indeed, a dangerous thing. If professional advice is not available but enquiry documents have to be compiled, the safest course is to leave the selected standard form of conditions unaltered (apart from completing any missing data as described in Section 4.11.4). Any special requirements are then added as special conditions of contract (Section 4.11.6) together with a statement that the special conditions take precedence over the standard conditions and shall prevail in the event of any conflict. Even this method, however, is by no means foolproof and can leave a lot of loopholes uncovered.

### 4.11.6 Special conditions of contract

1. Special conditions are new clauses to augment the general conditions of a standard form. Usually they deal with subjects not touched on by the standard form in which case the drafting requirement is to set down succinctly, but at the same time completely and unambiguously, the facts of the matter to be covered. Although, to this extent, it is often simpler to introduce a special condition than to amend a standard form condition, it must not be overlooked that the new clause will still have to operate consistently with what is said in the latter. This aspect must be duly considered after a new clause is written to ensure no conflict or ambiguity is being introduced.

Although the range of possible subjects for special conditions is large (see below for examples), in any one enquiry they will normally deal with the peculiarities of that one contract and one site. There is a choice of ways in which a new subject may be written into the enquiry documents:

- In the specification (if it is a technical matter, or in some way affecting one).
- As a special condition of contract.
- In the site regulations (if it deals with site administration or discipline).
- In separate correspondence (e.g. an Engineer's written instruction) if it is not essential to incorporate it into the formal contract.

The decision as to whether it should be a special condition rests on whether its breach should entitle the other party to cancel the contract and claim damages; if it is not so important a matter, it should be introduced in another way.

2.  The following are typical examples of subjects for special conditions; in some instances they are known to be partially covered in at least one of the more common standard forms of conditions, and this should be checked. In a contract in which they are of minor importance, they could rank as part of the technical specification or site regulations, as we have indicated above. Thus documentation might be an example of the former, and noise levels during erection of the latter.

- Special terms of payment (e.g. extended credit).
- Patents and licences: inventions arising.
- Applicable law.
- Official language for contract, notices and Engineer's communications.
- Obligatory use of local labour, materials, plant.
- Contract price adjustment and applicable formula and indices.
- Co-operation with other contractors on site.
- Insurance, special risks (Marine, Professional Indemnity, All-in Transit, 'Difference in Conditions', etc.).
- Fair wages to be paid.
- Union membership of work force.
- Prohibition of bribes and other unethical actions.
- Security, secrecy and Official Secrets Act.
- Protection and disposal of historic, valuable, archeological, etc., finds on site.

- Inoculation against endemic or epidemic diseases.
- Safety and welfare – the Health and Safety at Work Act 1974 – appointment of safety officer – contractor's site safety plan.
- Prohibition of access to named places.
- Restrictions on noise levels, dust, fouling of public highways, fire hazards, use on site of welding and flame-cutting equipments.
- Control of demolition work, explosives, etc.
- Responsibility for damage to public services: liaison with statutory authorities.
- Provision of facilities for sub-contractors and nominated sub-contractors.
- Record documentation to be provided. Operating and maintenance instructions. Number of copies. Standards.
- Liability to make spares available for a specified number of years, call off on demand.
- Inspection at manufacturer's works.
- Requirements for testing of plant performance.
- Responsibility for payment of local taxes, import duties, and arranging work permits.
- Applicability of the provisions of the Finance Act regarding payments of building labour net of tax.

3. A number of statutory regulations or acts may apply to the particular circumstances of the contract or the site. Although the contractor is often made responsible by law for recognizing and observing their provisions, there is virtue in reminding tenderers that they will apply, in order to:

- Remind the contractor of their existence and effect.
- Spell out the main influences on the contract.
- Confirm the responsibilities of the two parties respectively in regard thereto.
- Specify the method of fulfilling the responsibilities augmented by local site details.
- Limit any contingent liability falling on Employer.

The following lists a few of such acts, regulations and requirements:

- Official Secrets Act.
- Racial Discrimination Act.
- Finance Act 1971 (Employment of Persons).
- Health and Safety at Work etc. Act 1974.
- Control of Pollution Act 1974.

- Petroleum Consolidation Act 1928.
- The Mines and Quarries Acts 1954–1971.
- Code of Practice for reducing the exposure of Employed Persons to noise (HMSO 1972).
- BS Code of Practice for noise control on Construction and Demolition sites (CP 5228).
- Special requirements in relation to British Railways Board when the works are on or adjacent to bridges, tracks, embankments, etc.
- Various schedules of requirements imposed by statutory authorities (e.g. electricity, water, gas, gas-grid, etc.) when the works are on or adjacent to their systems.
- Special requirements in relation to British Waterways and river authorities when the works are on or adjacent to their territories.
- Regulations relating to the employment of foreigners (work permits, registration, routine reports, etc.).

### 4.11.7   Contract price adjustment clauses (CPA clauses)

1. A matter for decision by the Employer when drawing up an enquiry is whether the resulting contract should be subject to contract price adjustment. In times of inflationary uncertainty it can help both parties:

   - The contractor is protected (in good measure if not completely) against unpredictable cost increases, which he would otherwise have to cover in his tender price by sufficient contingency provision to meet the worst anticipated situation.
   - The Employer is protected against having to pay excessive contingencies in the contract price and from having to deal with a contractor who is striving to recoup a loss situation by skimping work, multiplying claims and, in the extreme case, appealing for *ex gratia* help to keep going on the contract.

   In making his decision, the Employer must weigh the foregoing advantages against those of a firm contract price for which he can, at the outset, make specific financial provision, and against which he can reliably measure progress during the contract period. With inflation around, say, 6% per annum he might well regard a 2–3 year contract period as his 'water-shed', but it must be remembered that in many countries the inflation rate is well into double figures, and for these a period of 12 months might reasonably be considered a maximum. The extent of the adjustment permissible to the contract price is usually specified by a formula in the CPA clause and is based on some recognized statistics or indices of ruling costs, such as those published from time to time by government departments, industrial associations or similar independent authorities for such a purpose. This safeguards the interests of both parties. Nowadays many countries (including all the major industrial ones) publish recognized official indices of costs ruling in their main industries. Although in the examples below we shall refer to those used in the UK, our comments could equally apply to the equivalent indices elsewhere. Those relating to a specific contract, of course, must reflect the costs the contractor has to meet in the countries from which he draws his labour, supervision, plant, machinery, materials, etc., for the execution of his contract. Though Employers entering contracts provisioned and run from abroad may not have their own local indices, they will be primarily concerned with those ruling in the original exporting countries.

Some standard conditions of contract, however, do not tie the permissible price adjustment to a formula but require the contractor to justify any claim for an increase. One such example is the IMechE/IEE Model Conditions Form 'A'. Justification can be laborious and very time consuming, especially with contracts not based on a detailed schedule of rates. It involves maintaining full records, not only to establish the rates and prices on which the tender price was based, but to show that increases in cost have, in fact, been experienced, their amount and (often most complex of all) the point of incidence when increases first reflected on the contract costs. Increases in overheads are difficult to substantiate, as are labour costs by any contractor who habitually pays his employees more than the standard nationally agreed rates. Potential sources of disagreement between the parties are many and it is advisable to rectify the position by an amendment to the clause introducing a suitable formula and indices.

Formulae almost invariably take the general form:

$$P_R = P_B \left[ A + \left( B \times \frac{L_R}{L_B} \right) + \left( C \times \frac{M_R}{M_B} \right) + \left( D \times \frac{N_R}{N_B} \right) \cdots \right]$$

where: $P_R$ = Adjusted contract price
$P_B$ = Original contract price
$A, B, C, D$, etc. = The fractions of the contract price represented by the cost of different services or commodities such as labour, transport, steel, copper, oil, etc. (The total $A + B + C$, etc., must, of course, equal 1.00). $A$ clearly represents a fixed part of the contract price.

$L_R$, $M_R$, $N_R$, etc. = The respective indices applicable at the date of adjustment.

$L_B$, $M_B$, $N_B$, etc. = The respective indices at the base date (used by the contractor in calculating his tender price). The base date can, with advantage, be quoted in the enquiry, e.g. 21 days before the date for return of tenders (see 5 below).

A number of appropriate formulae have been developed to suit different engineering disciplines, and have, by common acceptance, become widely recognized and used. We describe three below. Two of the better know were produced by Economic Development Committees in the early 1970s and published by the National Economic Development Office (NEDO). Originally they were familiarly known by the names of the chairmen of the two committees – 'Baxter' and 'Osborne' – but are today better known by the titles adopted for the industries by their National Joint Consultative Councils, namely:

- 'NCC Price Adjustment Formulae for Civil Engineering Contracts' and
- 'NCC Price Adjustment Formulae for Building Contracts'

The use of these formulae has been virtually standardized in the two industries and they warrant a more detailed description.

2. *NCC Price Adjustment Formulae for Civil Engineering Contracts.* These formulae are based on specially devised indices compiled statistically each month by the Property Services Agency of the Department of the Environment, and published by Her Majesty's Stationery Office (HMSO) in their *Monthly Bulletin of Construction Indices (Civil Engineering Works)*. They are also copied in various technical journals, usually during the fourth week of the month, and they can be seen on the television *Prestel* news service of British Telecommunications. The indices remain 'provisional' for three months from first publication, after which they are recalculated by the compilers and a firm figure issued. The indices represent:

   (a) The relative inclusive cost of employing labour and supervision in the civil engineering industry, i.e. current average wages plus all other incidental employment costs and loadings.
   (b) The relative cost to the contractor of providing, running, and maintaining constructional plant, engineering equipment and vehicles of patterns used in civil engineering works.
   (c) The relative market prices of nine groups of typical materials used in civil engineering works.

The procedure for applying the monthly indices to a price-adjustment formula has been set out in an explanatory booklet published by HMSO entitled *Price Adjustment Formulae for Civil Engineering Contracts*. It has also been incorporated as a detailed instruction by the Institution of Civil Engineers jointly with the Association of Consulting Engineers and the Federation of Civil Engineering Contractors in a model Special Condition designed to be added to their joint General Conditions of Contract (the so-called ICE Conditions of Contract) whenever the contracting parties decide that price adjustment shall apply to their own contract.

In applying the formula, the contract works are broken down and analysed at the enquiry stage, and costed into portions corresponding to the areas covered by the several indices. The division is therefore:

- A fixed (non-adjustable) portion currently standardized at 10% of the whole.
- Estimated labour and supervision costs.
- Estimated cost of providing, running and maintaining plant, equipment and vehicles.
- Estimated cost of each of the nine materials groups.

These portions are all expressed as a decimal part of the whole, but must of course, total 1.00.

The sub-division is usually done by the Employer prior to inviting tenders, and is based on his evaluation of the designs, drawings and bills of quantities he will issue with his enquiry. It reflects the Employer's view of how a normal contractor would carry out the contract, the plant he should employ, the temporary works he would need to erect, the average round-trip for spoil-disposal and the like. When tendering, a contractor will be using the same drawings and bills, and will usually accept the Employer's sub-division of the works. If, however, his plans for executing the works differ to a considerable degree from the Employer's concept, he may disagree with the figures and a mutually acceptable division has then to be decided as part of the tender negotiations.

Once a breakdown has been agreed and incorporated into the contract, it remains unchanged throughout the contract period regardless of how the composition of the actual work done may change or the contractors methods alter, as a result of remeasurement or changes introduced by contract variations. Insofar as the breakdown schedules for the various tenderers may differ following negotiations during the tender period, the anticipated effect on the total price the Employer may have to pay for the works needs to be considered during tender appraisal, in case the order to tenderers is affected by it (see Section 7.6.5 below).

Whenever a payment becomes due to the contractor under the contract, the figure for each item of the sub-division is adjusted to date for changes in its index. The total will then be 1.00 no longer but will have changed by a so-called 'Price Fluctuation Factor'. The amount due to the contractor (at the rates in the bills of quantity) is then adjusted by this same factor (up or down as the case may be).

Structural steelwork merits special mention since so much of its value is the cost of fabrication and finishing in a fabricators factory before it is delivered to the contractor for erection on site. It is already one of the nine classes of materials in the contract sub-division mentioned above, and when it forms only a small and simple part of the whole works the treatment it thereby receives is adequate for contract CPA calculations. If, however, fabricated structural steel forms a major (or the whole) part of the contract, the generalized formula no longer suffices. A separate NCC formula has been produced, based on DoE indices for:

(a) Cost of labour in fabrication and finishing of the steelwork, and also in the erection of the structures.
(b) Cost of structural grades of steel, rolled to section.

The process of calculating the CPA (which is quite complicated) has been incorporated into another Special Clause for adding to the ICE General Conditions of Contract. A decision to apply CPA to a proposed contract in the civil engineering field therefore calls for an initial appreciation of which is the most appropriate formula for the work concerned:

- The civil engineering formula.
- The structural steel formula.
- Both (each to its own part of the works).

If the structural steel formula is used, the tenderer must be required by the enquiry to state in his tender the rates he has used for

- The average cost per tonne for materials as delivered to the fabricator.
- The average rate per tonne for storage, handling and erection of the finished steelwork on site, including overheads and profits.

3. *NCC Price Adjustment Formula for Building Contracts.* A parallel formula to that used in civil engineering has been published and adopted for the building industry. Both are based on the same

principles and operate in a similar manner, but that for the building industry is considerably more complicated in its detailed application. The breakdown of the contract work is now divided into 49 groups of materials as well as five further divisions for specialist engineering installations commonly used in building works, namely electrical, heating and ventilating, catering equipment, lifts and elevators, and lastly structural steelwork.

As with the civil engineering formula, the indices corresponding to the items of the breakdown are compiled monthly by the Property Services Agency of the DoE, and published by HMSO in their *Monthly Bulletin of Construction Indices (Building Works)*. Indices remain provisional for the first three months and, after reassessment, then become firm. To assist application of the formulae, HMSO publish two booklets:

(a) *Price Adjustment Formulae for Building Contracts – Guide to application procedure*

and (b) *Price Adjustment Formulae for Building Contracts – Description of the Indices*

The Joint Contracts Tribunal (JCT) has adopted the NCC formulae for use with their joint Standard Form of Building Contract, published in a number of different forms by the Royal Institute of British Architects (RIBA). A clause written into the JCT Standard Form (Clause 31F in the 1963/1977 edition or clause 40 in the 1980 edition) calls up the formula for CPA: if that clause is not deleted, other methods of CPA are excluded. The same clause also calls up as part of the contract documentation the JCT Formula Rules, set out in their document *Standard Form of Building Contract – Formula Rules (1980)*. This document explains the rules and their method of application to a contract: a further explanatory document on the practical implementation of the formulae is the JCT Practice Note No. 18 *Adjustment of the Contract Sums by means of the Formulae*.

It is perhaps worth recording that whilst the JCT Standard Form of Building Contract is thus tied to its own Formula Rules, the ICE General Conditions of Contract can use either the NCC civil or building formulae (by suitably wording the Special Clause which needs to be incorporated). The use of the building formula will, however, be exceptional unless the works are largely or wholly composed of building operations. In other cases, building works can quite adequately be catered for by the civil engineering formulae.

4. *Electrical and mechanical plant contracts.* In view of the wide variety of products in these categories it is not surprising that there has been

less tendency in this field to standardize CPA formulae. Much more, the basic formula shown in 1 above has been adapted by manufacturers to reflect the predominant materials and grades of skilled labour involved in their individual processes. There will be a greater tendency for tenderers to call up their own formula to replace any general formula proposed in the enquiry documents.

This situation is reflected in the standard forms of conditions of contract common to this field. Probably the most used, the IMechE/IEE Model Form of Conditions 'A', includes a CPA clause (which has to be removed by amendment if CPA is not to apply) but makes no mention of any formula. If the clause is retained, this absence must be rectified by amendment before an enquiry using this standard form is issued. The standard form of conditions of the Institution of Chemical Engineers has no CPA clause as such and no formula is recommended.

One formula which is widely recognized, however, is the 'BEAMA' formula in which the number of variable terms is reduced to two – one for labour and one for materials. It is published by The British Electrical and Allied Manufacturers Association and uses as indices for materials those published monthly by the DTI in the *Trade and Industry Journal*, Table 1, either for Electrical Machinery or for Mechanical Engineering Materials. For labour it uses figures issued monthly by the BEAMA based on National Average Earnings Indices for the Engineering Industry produced by the Department of Employment. The BEAMA formula is in no way tied to the BEAMA standard forms of conditions or to purely electrical apparatus: it can be used with a wide range of contracts for the supply of plant of a mechanical or electrical (or mixed) nature, with or without erection. BEAMA also publish a number of variants of the same formula using factors, constants and indices appropriate to different types of plant, e.g. turbo-generation plant or electronic equipment, and for export purposes.

Some formulae (including the BEAMA) use as the 'factor applicable at the date of revision' an average of the monthly indices over a part of the contract period. They also allow, for example, for the fact that in plant manufacture, materials are purchased early in the contract period, whilst labour is mostly involved in the later stages of the contract period after the materials are in and the job can be released to the shop floor.

A closely parallel set of formulae, useful in contracts dealing with pipe work and similar steel-based products, are the three versions published by the Water Tube Boilermakers Association. These again employ only two variable factors, materials (usually based on the DTI monthly figures of Indices Nos 311 and 312 (Table 2) from the *Trade and Industry Journal* mentioned above) and labour (based on WTBA's own index which in its turn is based on the labour rates and allowances nationally accepted by the Engineering Employers Federation). The three versions referred to have different values of *A, B, C* (shown by the formula in 1 above) designed to apply to home contracts including erection, home contracts excluding erection and overseas contracts respectively. The Water-Tube Boilermakers formulae are of particular interest as they include a method of applying CPA to contracts for the supply of a number of identical plants (such as boilers or other machines) in succession over a considerable number of years. Such a 'repeat' contract raises a number of problems when devising an equitable formula at the time of tendering, the formula acting unfairly when applied to the later units if a single base-date is used throughout the contract. The WTBA system establishes for each unit to be supplied, a revised basic contract price and a new base date, both related to the start of work on the new unit.

5. *Influence of choice of reference dates.* Finally reference must be made to the influence of choice of dates on the amount of CPA calculated by any given formula. There are two aspects, both resulting from the fact that the indices for labour and materials are usually published and come into effect on fixed days each month. In a period of continued inflation, the indices can be expected to step up each month on a known date, and when inflation rate is high each step can be appreciable.

The first conclusion therefore is that, to allow a direct comparison between tenders, all must be priced as at the same base date. This date is preferably specified by the Employer in his enquiry and can, for example, be the date set for the return of tenders, or a stated number of days prior thereto. The second point is that by subtle selection of the delivery, completion or invoice dates (whichever determines the upper index when applying the relevant formula) so as to embrace – even by 24 hours – a new index figure, a contractor stands to gain an additional CPA sum if the index has gone up on the month. This has to be taken into consideration when making tender appraisals, as it can easily tip the balance between two apparently equal tender offers. Put in a rather different way, if two tenders have the same tender price, the one taking longer to complete will probably, in an inflation climate, cost the Employer more in the end as a result of applying the CPA clause. With high inflation rates the 'longer period' may be as little as one day!

## 4.12    THE GENERAL DESCRIPTION OF THE WORKS

**4.12.1**    The general description introduces the scope and purpose of the contract works and co-ordinates the various contract documents such as drawings, technical specifications, site data and the like.

It will vary considerably in size with the complexity of the works. In a contract for the provision of simple plant a general description may not appear at all, its purpose being adequately covered by the invitation to tender ('you are invited to tender for the supply of . . . . . . . . . . . .'). In many contracts it will comprise only a few paragraphs. In the more complex contracts, however, it can become a sizeable document in which the works are broken down into sections each with its schedule of technical documentation and drawings. The main purpose of the general description is then to co-ordinate the functions of the sections and to specify their interfaces. It might, for example, include:

- Statement of purpose and scope of the Employer's project.
- The broad project timetable.
- Definition of the works.
- The project site, location (with map references), access, traffic routes, contract site, site layout, site maps, drawings, etc.
- Site topography, geology (if not in a site-data document).
- The sections of the works comprising the contract, their respective scopes, layout, interfaces. Operational interrelation.
- Completion priorities and interdependence during construction of the various sections.
- Methods to be applied to the control of progress of the works.
- Expected performance of the works as a whole and any combined performance tests to be made on completion.
- Standard specifications, regulations, etc., which are to apply to all sections of the works.
- Schedules and descriptions of the technical contract documents which treat the various sections of the works.

**4.12.2**    The technical documents for a large contract are produced by a number of different people working concurrently and it is inevitable that an occasional discrepancy or conflict will occur between them. It is not unusual therefore, in such contracts, for the general description to prescribe in a schedule which document shall prevail over which, and to specify the action a tenderer shall take on discovering an error. He must be prevented from making his own guess at the correct interpretation and telling nobody.

**4.12.3** Some technical matters involve contractual or legal aspects which are rightly dealt with by clauses in the contract conditions, special conditions or site regulations. They may also have been considered by the writers of the technical documentation which should be checked that it neither duplicates nor conflicts with the contractual form. Such points might, for example, include:

- Procedure for introducing variations to the works specification and the pricing thereof.
- The issuing of certificates of various types by the Engineer.
- Schedule of the various tests to be made on the works (see Section 4.14.5) and the implications of failing to pass them.
- Handling of nominated sub-contractors' activities.
- Responsibility for care and safety of materials, plant and the works – insurance matters.
- Delay in completion of the works or sections of the works.
- Responsibility for fencing, lighting and watching of the works.
- Terms of warranty and/or guarantees affecting the contract.
- Pollution, noise and other local nuisances.
- Use and maintenance of site roads, parking of vehicles, storage sites for materials, etc.
- Access for and assistance to other contractors concurrently on site.
- Rights of way, safety of the public.
- Wearing of protective clothing whilst working on site.
- Responsibilities in regard to local authorities, public services, railways, electricity and gas grids etc., affected by the works.

## 4.13 SITE DATA AND SITE FACILITIES

The conditions at the site can affect the cost to the contractor of executing the works in two ways:

- The natural and climatic conditions to be expected at the site; man-made obstructions (buildings, pipelines, utility installations and the like) existing on the site, i.e. site data.
- The facilities which the Employer undertakes to provide at the site for use by the contractor, i.e. site facilities.

This information must therefore be presented to the tenderer at the enquiry stage. With some contracts (notably constructional work) the whole method of execution may depend on the site conditions and the information is presented as a separate document. With others (notably the supply of plant on to prepared foundations) the site conditions may have little importance and the few it is necessary to mention may be incorporated in the technical specification or in the site regulations.

### 4.13.1 Site data

The following are some of the subjects which fall to be dealt with under this heading. Information must be presented as 'factual' data, quoting the source from which it has been obtained and, if possible, the time at or period over which it was recorded. The source gives an indication of its possible reliability and the times may embrace periods of abnormal conditions. For example the measurement of water-table level which happened to be made at the end of a prolonged dry spell might give a quite erroneous impression of typical levels.

No deductions from the data presented should be quoted by the Employer/Engineer. Under normal circumstances the contractor is solely responsible for the execution of the works and the methods of construction he employs: he must be responsible for ascertaining the site conditions. Factual information may be given him to save time, but he must himself decide on its implications and the effect these have on his operations. Typical subjects, of which the following are examples, may be grouped under two headings:

(a) *Recorded results of a site survey made for the purposes of the project:*
   - Nature of the ground (topsoil, substrata, outcrops, etc.).
   - Borehole investigation data.
   - Water table: availability of water (potable and for industrial uses).
   - Ruling gradients: contoured maps and site-plans.
   - Natural obstructions (woods, ponds, escarpments, ditches).
   - Man-made obstructions (buildings, embankments, drains, overhead power lines, etc.).
   - Public service distribution systems affecting the site.
   - Access roads and tracks. Public highways. Rights of way. Maximum load routes. Turning places and circles.
   - Accessability of site from port/airfield suitable for freight. Rail facilities to the site or the locality. Load-handling facilities: cranage, transfer machines.

(b) *Published data from various sources (more usually provided in respect of sites overseas):*
   - Local geology.
   - Altitude.

- Rainfall, snowfall, liability to flooding (in all cases stating period of year and extent) – monsoon period.
- Temperature variations (diurnal, month by month) – extremes.
- Humidity range (diurnal, month by month).
- Prevailing wind, liability to storms, maximum wind force. Occurrence of sandstorms, drifting sand.
- Liability to earth tremors, earthquakes.
- Local health conditions, endemic or epidemic diseases.
- Tide tables.

### 4.13.2  Site facilities available to contractors

Arrangements made available on site by the Employer/Engineer which the contractor may take advantage of when executing the contract. In some cases the enquiry will include pro formae on which the contractor is required to record the extent of his proposed usage (see Section 4.9).

Typical subjects include:

- Access to site by road/rail for workpeople/materials/plant.
- Restrictions on loads or types of traffic.
- Employer's plant available for use (cranes, tractors, pumps, trucks, etc.) with types, sizes and conditions of usage. Provision of operators. Vehicle washing points.
- Supplies of electricity, gas, water, compressed air, potable water available for use. Pump petrol and Derv.
- Telephone and telex facilities: allocation to contractor and restrictions on use. Mail reception and despatch facilities.
- Negotiated wayleaves.
- Vehicle parks, hard standings. Vehicle unloading facilities.
- Storage space for materials, plant: covered, open air.
- Huts/offices, and with what facilities.
- Canteen facilities, vending machines, meals on wheels, facilities for tea making. Rest rooms.
- Lavatories, ablution facilities, baths available for use by staff/workpeople.
- First-aid, medical centre, ambulance service, local hospitals, facilities for inoculations/vaccinations.
- Hygiene facilities, pest control.
- Refuse disposal (burnable, soft and hard-fill rubble).
- Fire-fighting facilities and equipment. Alarm systems. Liaison with local services.
- Local labour – availability, quality and rates.

- Labour camps, social centres, laundries, cinemas, radio and TV services. Arrangements for the reception of wives and families: schools.
- Local sources of raw materials, quarries, cement plants, etc.

In all cases, under site facilities, it must be made clear in the enquiry, whether the facility is provided free or is charged and if the latter, the method of assessment of the contractor's usage and the rate of charge to be levied. It may be necessary in some cases to say whether the contractor is at liberty to provide his own facility and ignore that provided by the Employer.

## 4.14   THE TECHNICAL SPECIFICATION

**4.14.1**   The Technical Specification is the detailed exposition of the Employer's requirements in respect of the contract works. It will clearly vary in size and content just as widely as engineering contracts themselves can vary, so that we can do little more in this book than make some general observations. This however, does not mean they are unimportant.

In civil engineering contracts the Technical Specification, backed by the contract drawings and the bills of quantities, customarily provides a fully detailed description of the permanent contract works in all their facets.

With building contracts (such as those normally executed under the JCT Standard Form of Building Contract) it is customary to take off bills of quantities in considerably finer detail than with civil engineering works so that, with the contract drawings, they fully describe the building works concerned and no separate technical specification is necessary. Indeed the JCT Standard Form of Contract (with quantities) nowhere refers to a 'specification'.

At the other extreme, in contracts for the supply of mechanical or electrical plant, bills of quantities as such are frequently non-existent and the technical specification is largely functional in content. It is left to the tenderer using his specialist knowledge and manufacturing expertise to submit his proposals as to how the Employer's requirements can best be met.

Whatever the nature of the works, the technical specification, backed by the drawings and bills of quantities, must say everything the Employer wishes to lay down about the form the works must take. The contractor will only be bound to carry out in a good professional manner what the contract says. He has no obligation to provide unstated features which the Employer had firmly in mind but unfortunately forgot to specify.

Especially with plant contracts, after-thoughts are apt to conflict with the contractor's plans and designs already made, involving expensive and wasteful redesign and reworking.

**4.14.2** Considerable thought must be given to the layout of technical specifications. All too frequently they appear as jungles of data following no recognizable format. Each subject should, whenever possible, be covered completely in one place, and be readily locatable by an index. If this is not done the tenderer can never be sure he has read the whole specification for any item.

The technical requirements should therefore be carefully fitted to a well-planned skeletal framework:

- The technical specification is divided into a number of main sections each with a main heading. The sections may well be 'sections of the works'.
- Each section is divided into a number of relevant subjects each with its sub-heading.
- Each subject is then divided into a number of relevant features which need to be specified, each with its sub-sub-heading.

Only when this framework is finished should the technical material be drafted on to it, each subject being dealt with under its appropriate sub-heading. An index or table of contents (i.e. a list of headings and sub-headings) is then added to complete it.

The works to which the enquiry relates are divided in this way into self-contained portions, each portion being allotted a main section, inside which that portion is specified completely.

Matters having a general application to many (or all) of the portions of the works (for example standard design specifications, BSI publications, wiring regulations, codes of practice) are introduced together into a 'general' section to form, as it were, the ground-rules for the works. In the same way, a main section may be devoted to a schedule of the various inspections and tests the works are required to undergo (see Section 4.14.5). In both cases the particular aspects which affect any one portion of the works are dealt with in the section of the technical specification relating to that portion, cross-references to the general section being made as necessary to avoid repetition.

If it is found necessary to include copies of less-well-known standard documents with the technical specification, these should be kept as annexures and on no account included in the body of the technical specification itself. The annexures may even form a separate volume of the enquiry documents.

By adopting such a division of the technical specification, not only is the amount of 'page-hopping' necessary to understand a given subject reduced to a minimum, but each feature can be readily located (by the Engineer and the tenderer, as well as the writer thereof) and the completeness of its contents can be systematically checked.

The British Standards Institution publication PD612 (May 1967) *Guide to the Preparation of Specifications* may give further useful guidance and help in standardizing the format of technical specifications.

**4.14.3** Due largely to the limitless variety and complexity which 'plant' can take, the documentation for plant contracts is much less standardized than in the construction industries. There is no equivalent (for example) to the standard methods of measurement applicable to the latter, and the detailed bills of quantities they give rise to (see Section 4.18.2).

With plant, the technical specification, as indicated above, is often restricted to prescribing the properties and performance the Employer requires, the operating conditions the plant has to meet, and such dimensions as are necessary to enable it to fit into place in its location, and interface with any other plant with which it has to operate. The tenderer is therefore restricted only in so far as is necessary, and remains free to apply his experience, normal practice and preferred methods in arriving at his proposals. Competition between tenderers is not on a basis of price and quality only but the ingenuity and effectiveness of the tenderer's design become important factors.

Such a plant specification must require the tenderer to produce with his tender sufficient evidence to expose the features of design and convince the Engineer that the product put forward will meet the Employer's requirements. The evidence might involve the submission of:

- A formal specification.
- Design calculations.
- Drawings of construction and layout.
- Dimensions, weights and materials.
- Automatic controls and safety measures provided.
- Descriptions of sub-assemblies and parts (to indicate source and quality).
- Reliability in service – mean time between failures (anticipated, or from past experience).
- Test reports on prototypes or existing installations.
- Performance figures achieved on earlier installations.

The technical specification for plant will also generally call on tenderers to quote anticipated or guaranteed operational data for the plant they propose, including:

- Maximum rated output and product purity.
- Power consumption (i.e. efficiency) at rated full load, on maximum acceptable overload, and at specified part loads.
- Staffing and control required in operation.
- Consumption of lubricants, cooling water, etc.
- Noise and vibration levels in operation at full load.
- Radio interference levels generated.
- Running temperatures.
- Typical exhaust gas analysis.
- Routine maintenance tasks (and frequency).
- Recommended holding of running spares for (say) 12 months.
- Expected life of wearing parts.

**4.14.4** From the points of view of interchangeability, repair and maintenance there are advantages to the Employer in standardizing throughout his undertaking the makes, types and – to some extent – the sizes of common equipment (such as plant instrumentation, drive motors, pumps, etc.). This should be borne in mind when drafting technical specifications, especially for sites where the Employer already has installations in operation.

**4.14.5** An important item for inclusion in all technical specifications is a full description of inspections and tests which the Engineer intends shall be applied to the works, and the standards to be achieved in them before the works will be regarded as acceptable and the contract to have been completed.

The problem with most civil and building construction contracts is not usually complicated, the criterion being whether the quality of materials and workmanship and the dimensions of each part of the works are in strict accordance with what was specified in the designs of the Employer/ Engineer. Acceptance usually rests therefore on constant vigilance and checks being applied throughout the construction phase by the Engineer's representatives on the site followed by a final inspection when work is claimed to be complete to ensure nothing has been omitted. The finished works must match the Employer's designs.

With plant contracts on the other hand, the problem is much more involved since most of the work is done off-site in the factories of the contractor and his numerous sub-contractors. The detailed design was not made by the Employer and the manufacturing drawings and specifications are not available to him. The performance of the plant cannot be checked item by item but has ultimately to wait final installation on site. It is therefore necessary to provide for a series of inspections and checks at stages throughout the contract to ensure as far as possible that a satisfactory product is going to result. The technical specification must schedule these tests and include test procedures, measurements and the interpretation thereof, as well as laying down the minimum results which will be considered satisfactory. This is all required in considerable detail if there is to be no room for dispute during or after a test, as to whether the item submitted to it has or has not passed. Plant tests and inspections will usually comprise most or all of the following.

*Inspection and test-checks* of materials and components: the workmanship, type and standard of finish and performance of subassemblies at all stages of manufacture. They may include a right of examination by the Engineer of the more important works drawings and specifications being used by the contractor, and extend into the premises of the more important sub-contractors. Sub-assemblies may be subjected to vibration or environmental tests, reliability runs, or performance checks with limit-value services.

*Factory acceptance tests* are specified so as to demonstrate that the plant can be properly assembled and functions satisfactorily as a free-standing equipment. Where possible some partial or artificial loading should be applied, but in some cases it will be feasible (owing to the nature of the plant) to test it at the factory on 'no-load' only. Manufacturers normally apply a routine set of tests to all their standard products, and for such cases approval or otherwise by the Engineer of their test schedule and test reports may be all that is necessary.

*Site acceptance test (or test on installation)* is used here to mean an intermediate test which is sometimes included if a long interval is expected between delivery and erection, and completion of the project to a stage when full tests on completion (under actual load conditions) can be applied. It is especially useful for installations overseas, to enable the installation team to be cleared and withdrawn from site. Site acceptance tests (and tests on installation) are a repeat of the factory acceptance tests to expose any damage or deterioration which might have occurred during shipment or prolonged storage prior to installation on site. This can then be rectified and delay is avoided when tests on completion come to be made.

*Tests on completion.* The test schedule must be such as to check the plant against all the requirements of the technical specification. All tests must be made with the plant operating on its designed load and prescribed overload, and preferably in full conjunction with its associated plant (if this is available) as an intrinsic part of the project. The tests are usually performed by the contractor under the supervision of the Engineer, the appropriate raw materials being provided by the Employer free of charge to the contractor.

*Performance tests* are made to establish whether the plant meets the standard of performance laid down in the contract in respect of maximum throughput, power consumption and efficiency, quality of product and the like. They are usually made after the plant has been taken over by the Employer and become operational, when it is allowed a suitable running-in period under the eye of the contractor prior to its performance check. The plant is operated by the Employer but the contractor is usually allowed to have the plant operated to his satisfaction during the progress of the tests. The condition of contract which provides for the performance tests to be made, must also prescribe the penalties payable by the contractor for failure to achieve the required performance.

**4.14.6** The conditions of contract should contain a clause giving the Engineer the right:

(a) To reject any materials components or sub-assemblies intended for use in the plant which, on inspection, *are in his opinion* of inadequate design, improper quality or sub-standard workmanship; and

(b) To refuse to accept delivery of the plant until it has passed the prescribed factory acceptance tests and is manifestly in a suitable condition to warrant its leaving the factory. This is important partly because it is common practice to allow payment of a substantial part of the contract price when delivery is made and partly because it is much more difficult to rectify faults and install replacement parts once the plant is away from the resources of the factory and the engineers who designed it.

**4.14.7** The tests on completion are the main proving of the plant and are therefore made where possible under the intended full operational loading. They usually include a short run at a reasonable percentage above full rating, as a safety measure. The test specification must ensure that all the operational parameters are checked including environmental aspects such as noise, vibration, smoke, smell, fumes and effluent at various loadings. All automatic controls and safety measures built into the plant must be caused to operate without failure a specified number of times on random occasions, and the plant shown to restart satisfactorily on command thereafter. On completing the test schedule it must have been demonstrated that the plant is satisfactory in every respect (apart from minor points which are recorded and are made good by the contractor immediately following the tests). It is then taken over by the Engineer on behalf of the Employer who is thereafter responsible for operating it under whatever supervision or instruction by the contractor is provided in the contract. The period of maintenance under warranty usually starts at this point.

**4.14.8** The performance tests aim to determine the standard of efficiency which the plant can achieve in operation, and shortcomings during the tests will usually be a matter of degree above or below a promised level of performance. Whilst total rejection of the plant for low achievement can be written into the contract conditions (in spite of its acceptance and taking over previously), this usually acts against the best interests of the Employer, owing to the inevitable delay in supplying a substitute. The penalty on the contractor is therefore usually agreed as a scale of heavy liquidated damages calculated to represent the fact that any reduction in performance inevitably means higher operating costs to the Employer (or an inferior product of lower worth) over the whole life of the plant. It is not unusual for the parties to agree, on the other hand, a bonus payment to the contractor if the recorded performance is markedly superior to that promised in the contract.

The specification for the performance test is a difficult one to draft effectively since we are concerned usually with precise measurement of differences in performance of as little as 1 per cent or less. As the results can have serious financial consequences for the contractor, the Engineer must ensure that the tests and their interpretation have no cause for error or doubt. Particular care must be taken with sensitive parameters. For example in the drying of moisture from a slurry, small changes in the moisture content (representing a miniscule part of the total slurry weight) can require a considerable increase in energy consumption for the additional evaporation. Close control of this moisture content would therefore become a key requirement during the performance tests of such a plant.

On many varieties of plant a representative performance test can be fairly carried out only after the plant has had a chance to run itself in and settle down following successful tests on completion. Although the plant has been taken over and is operated by the Employer it is usual to allow the contractor to make any reasonable adjustments or modifications which, whilst not bringing the plant outside the terms of the contract, are likely to bring about an improvement in its performance. The Engineer will normally reserve the right to approve such changes beforehand and to supervise the contractor whilst putting them into effect.

The following is a check list of some of the aspects of performance tests which need to be dealt with in the technical specification:

- Interval after taking over the plant before test is made.
- Duration of test runs and hours per day.
- Continuous or intermittent runs?
- Any prescribed warming up period before test run is started?
- Levels of plant operation at which tests are to be made.
- Control of quality of raw materials to be treated (e.g. purity, moisture content).
- Control of operating supplies (e.g. calorific value of fuel, electricity voltage, cooling water flow and temperature).
- What parameters are to be measured and at what intervals?
- What instrumentation is to be used, who provides it and how is its accuracy independently verified?
- What rights does the contractor have in operating the plant or directing how it shall be operated during the tests?
- Can contractor adjust plant settings during a test or must the test be restarted?
- Who controls the tests, supervises their progress and maintains records?
- Who reads instruments and records instrument readings?
- Are contractor/Employer entitled to witness all tests? To check records? To query readings? To query interpretation of results? Or is Engineer sole authority?
- Are any maintenance/running repairs permitted during test period? During intervals between test periods?
- Can contractor elect to discontinue a test and start again without penalty?
- Can any forms of interruption or maloperation be ignored, and on what basis (e.g. drop in electrical voltage)?
- What test criteria are to be determined and who calculates them?
- How many attempts may be made at the tests by the contractor before damages are imposed and to what extent may he modify the plant in between tests?

The consequences of failure to reach the contractual level of performance and the damages payable (as well as bonus, if any, for better performance) are matters for the conditions of contract and not the technical specification, which restricts itself to the *modus operandi* of the tests themselves.

## 4.15   THE DRAWINGS LIST

Drawings form an intrinsic part of the technical documentation of the enquiry. They must be identified, listed and incorporated into the declaration of the form of tender. The Drawings List is often a part of the technical specification and gives for each drawing:

- Its reference number.
- Its revision letter.
- Its title.

The office of origin should be clearly shown on the drawing if it is not implicit in the reference number.

Wherever possible enquiry drawings should be:

- Black on white.
- Standard size (one – or at most two – of the A1–A4 range).
- Given a reference number unique for the project, whatever the source of the drawing.
- Folded to a convenient size (A4) to show the title-block and reference number.
- Not reproducible (to avoid unrecorded copies and the resulting amendment difficulties).

## 4.16   SCHEDULES OF NOMINATED SUB-CONTRACTORS OR PREFERRED SUB-CONTRACTORS

The Employer/Engineer may wish to designate in the enquiry that certain parts of the works are to be executed by the contractor through nominated or preferred sub-contractors. In either case it is customary to draw tenderers' attention to the fact in the instructions to tenderers, and to schedule details of the proposals in the enquiry documentation (often in the technical specification).

### 4.16.1   Nominated sub-contracts

It is generally true that the nominated sub-contract concept has arisen in connection with contracts in which the Employer retains responsibility for detailed design of the works. There are advantages to him in discussing direct with suppliers the design of sophisticated or critical equipment, or its appearance to give the architectural effect he wants. At the same time he negotiates a contract for the supply of the material he wants.

Whoever the main tenderer might turn out to be, he would have to sub-contract the equipment ultimately and by operating direct the Employer removes the items from the competitive and second-hand activities of the main tenderers. Direct negotiation in this way can also be of help where delays in design or provisioning require work to be put in hand in advance of the appointment of the main contractor.

The basic concept is that the Employer completes the contract arrangements for the sub-contract, nominates his chosen supplier to the main contractor who is then required, by the terms of his own contract, to embody the Employer's terms in an order he negotiates with the nominated supplier. In this way, the main contractor retains responsibility for the whole project including the sub-contract concerned. It is a procedure fraught with contractual difficulties, and these we give more consideration in Section 8.2.

It will be clear that the introduction of nominated sub-contracts has to be defined and imposed on a main contractor by clauses in his contract, since the concept itself has no intrinsic status at law. This is done, for example, in the ICE General Conditions of Contract (5th edition) at Clauses 58–59C. The idea offers little advantage (if any) in contracts for the supply of plant, where the Employer normally restricts his specification to functional requirements, and it is not recognized by the general conditions of contract which apply specifically to that field.

The estimated cost of each nominated sub-contract is calculated by the Employer/Engineer and included in the bills of quantities for the main enquiry as a separate entry – a *prime cost item* – so that it forms a firm part of the main contract price. The prime cost has three components, which are customarily included as three separate entries in the bills:

- The price actually paid by the main contractor to the nominated sub-contractor (less any sums which might arise from his own fault or delay) net of all discounts except that for prompt payment to the sub-contractor.
- Payment for the main contractor's own work in managing, supervising and controlling the nominated sub-contract, and providing direct assistance to the sub-contractor. This sum is either negotiated as part of the main contract, or otherwise is determined by the Engineer.
- A sum in respect of profit to the main contractor, and in aid of any other expenses he may incur. This is usually negotiated as an agreed percentage of the sum under the first item, and is so specified in the main contract.

If, when the sub-contract work has been completed, the actual sums differ from the prime cost entries written into the bills of quantities at the time of tendering, the latter are adjusted to the actual figures by an Engineer's variation order as provided in the contract conditions. It should be noted that, unlike any *provisional items* included in the bills, prime cost items are a firm part of the contract from the outset, and the contractor is entitled to receive payment for them in full.

In the enquiry documents for the main contract it is usual to give a separate schedule of nominated sub-contractors with whom the selected tenderer will be required to deal. The schedule is named in the Form of Tender so that the tenderer's acceptance of the details it contains and his undertaking, are both embodied in his offer and thus, if he is selected, in his contract. For each nominated sub-contractor the schedule should state:

- The work covered by the sub-contract.
- The name and address of the nominated sub-contractor (if known at the enquiry stage).
- The position reached by the Employer in his negotiations.
- Any sub-contract programme dates agreed.
- The item-numbers in the bills of quantities where the prime costs are included.

### 4.16.2   Preferred sub-contractors

To avoid the dangers and pitfalls of nominated sub-contracting the Employer can, as an alternative, specify in his enquiry a number of *preferred sub-contractors* for stated portions of the works. From among these he expects the tenderer to select and nominate one of his own choosing, and to employ him if he is awarded the contract. Such list of preferred sub-contractors is usually introduced as part of the technical specification, and may be accompanied by a note as to whether each firm has been approached and has indicated its willingness to do the work if selected. The tenderer is not obliged to choose in his tender any of the preferred firms for the portion of the works concerned, but if he does not, he is expected to give a very good reason. If his reason is considered insufficiently important by the Engineer it could result in his tender being rejected. Such a risk should be clearly stated in the instructions to tenderers, or in the schedule of preferred sub-contractors itself. A prudent tenderer would, of course, get his non-conformist proposal approved in principle before basing his tender on it.

Although not connected with the civil and constructional industries, it is not uncommon for the Employer to specify preferred sub-contractors (or more probably 'preferred suppliers') in enquiries for the supply of plant, but it would be unusual for him to put in hand any negotiations with the sub-contractor himself other than such technical discussions as were necessary to ensure he was able to provide what the Employer wanted. Reasons for preferred sub-contractors in plant contracts vary considerably: it might be that a system has to be designed around the characteristics of a key assembly which therefore had to be firm and fixed before the design could proceed. On the other hand it could be nothing more esoteric than a desire by the Employer to standardize on one make of electric motor, or one brand of instrument throughout his factory, and thereby to ease his spares and maintenance problems.

## 4.17  SITE REGULATIONS

Site Regulations embody matters which the Employer/Engineer wish to impose on contractors (and on their own employees) to ensure effective co-ordination and administration of the joint activities on the project site. Any rules, infractions of which are serious enough to be regarded as breaches of contract by a contractor possibly entailing a penalty of cancellation of his contract, must be dealt with as special conditions of contract (see Section 4.11.6) and not as site regulations.

The subject matter of site regulations may stem largely from standard sets of rules drawn up by the Employer for the control and protection of existing facilities, and the safety and welfare of persons employed at all his sites and factories. To these the Engineer may have to add further items more particular to the project site itself and the actual conditions which will prevail there during the course of the execution of the contract.

The following considerable list gives some typical items for possible inclusion in site regulations:

- Access routes to and from the public highway system.
- Site entrances, exits, traffic-flow systems, traffic control.
- Limits on weights/dimensions of loaded vehicles using roads.
- Limits on total tonnage transported by road daily.
- Access by rail: use and control of Employer's sidings and loading bays. Special safety measures to be observed.
- Rights of way for the public.
- Fouling of highways: wheel washing.
- Vehicle parking areas allotted to contractor: others expressly prohibited.
- Authority required to remove loaded vehicles from site.

- Examination of vehicles leaving site and their loads by Employer's site police.
- Site security arrangements, measures and checks.
- Site police arrangements: dog patrols.
- Personnel identification – procedures, passes, check points.
- Work permits, visitors' permits.
- Restricted areas: areas out of bounds to all contractors' vehicles/employees.
- Protection of the public – screens, guard rails, watching, lighting. Public observation stations.
- Anti-sabotage measures, alarms, actions.
- Fire precautions, procedures, alarms, location of fire points, liaison with works/town fire brigades, emergency action by contractors.
- Restrictions on storage of gas or inflammable fluids.
- Use and storage of radioactive materials.
- Regulations re blasting and use and storage of explosives.
- Site safety officer – coordination of safety-at-work measures.
- Prohibition of introduction on to the site of noxious, poisonous or objectionable substances.
- Rules to be observed re naked lights, unguarded fires, smoking or welding.
- Hygiene and sanitation arrangements for contractors employees/staff.
- Pollution of site, blocking or excessive use of streams, ditches, drains, sewers. Waste-water discharge points: grease traps.
- Safety in use of electricity, portable machinery, fires.
- Damage to Employer's plant, pipework, roads, etc.
- Dumping sites for excavated earth, demolished masonry, rubble and contractor's waste. Control of bonfires for contractor's waste.
- Cleanliness of site – removal of surplus materials, spoil, rubbish.
- Use by contractor's employees of Employer's canteens, rest-rooms, sports' facilities. Hours of opening.
- Restriction on hours and methods of working. Authorization in advance of work on site out of normal hours.
- Limitations on work at night – floodlighting, noise.
- Restrictions on levels of noise, pollution, etc.
- Avoidance of damage to or interference with site services (gas, water, sewerage, etc.). Location of local liaison officers of public services.
- Storage areas and buildings allotted for contractor's use.
- Local authority by-laws or instructions in force at site.
- Other contractors at work in vicinity – access permitted across site.
- Erection of hoardings, notice-boards, direction signs, etc.
- Photography, advertising, etc., on site.

- Holding of political or other meetings on site, distribution of leaflets, holding of demonstrations.
- Observance of local traditions, holidays, *tabus* or religious observances.
- Statement of charges (if any) and methods of calculation for use by the contractor of the Employer's site facilities.
- Nearest large town, its social facilities and means of access to and from site.
- Additional regulations applicable to site labour camp (if any).

## 4.18  BILLS OF QUANTITIES (OR SCHEDULE OF RATES)

**4.18.1**  Contracts based on bills of quantities or schedules of rates are in most cases those in the civil engineering and building industries where their use is standard practice. It is, indeed, the logical extension of a procedure in which the Employer/Engineer is responsible for the detailed design of the works that, having drawn up the design, he should analyse it into its different tasks, materials and items of work for presentation to the tenderers, one of whom will have to carry it out.

It will be realized that a tenderer is usually presented with an enquiry for a project with which he is entirely unfamiliar, and given a very limited time in which to estimate his costs and prepare his offer. This only becomes a possible task, if those already versed in the design detail have prepared an analysis for him, the bills of quantities, in a form he can price largely as an exercise in arithmetic, based on his experience in the industry.

The purposes for which bills of quantities are used are therefore:

- To assist in defining the works in detail.
- To enable a tenderer to price an enquiry rapidly and accurately.
- To analyse the works and ensure that no items or procedures are left unpriced or sources of cost or expense to the contractor are omitted.
- To group parts of the works into separate entities as required by the Employer/Engineer's cost-control systems.
- To establish for each item a unit rate of charge which can be used for pricing variation orders, the calculation of the actual cost of re-measured work and (if applicable) the calculation of any contract price variation which may become due to the contractor.
- To isolate from the constructional work proper, the cost of all 'once-off' or overhead forms of charges (e.g. site offices, supervision and management, insurance, licences, wayleaves, etc.).
- To make financial provision for nominated sub-contracts (prime cost items).

- To make financial provision for additions to or variations of the contract works not yet fully defined or adopted (provisional sums).
- To arrive at tenders in a form which can be rapidly analysed, compared and appraised bearing in mind that differences between tender prices will reflect tenderers methods, skills and organization – the technical content of the works being the same for all.

**4.18.2** The degree of detail to which the works are broken down and analysed in bills of quantities varies considerably. It can reflect for example the normal practices of the Employer/Engineer, the extent and variety of variation orders which are anticipated in the contract, the fineness of sub-division of the cost-control system which is to be applied to the contract. It can also depend on the extent to which descriptions in the bills of quantities are relied on to specify and define the items. In civil engineering contracts the definition is usually covered by the technical specification and the drawings, the bills being less detailed than in the building industry. Here, by contrast, it is customary to define each item by the drawings and the bills of quantities, there being no separate technical specification (the JCT Standard Form of Building Contract (with quantities) at no point mentions a 'technical specification').

To rationalize the systematic production of bills of quantities, their recognition and pricing by the tenderer, and their analysis and interpretation by the Employer and the Engineer, standard rules have been introduced and established by common usage.

Two such standards are widely recognized:

(a) *for civil engineering works*
*Civil Engineering Standard Method of Measurement* (1976), published by the Institution of Civil Engineers.
(b) *for building works*
*SSM 6 – Standard Method of Measurement of Building Works* (6th Edition, 1979), published jointly by the Royal Institution of Chartered Surveyors and the National Federation of Building Trades Employers.

These standards lay down among other things, the order in which items are to be listed and the way in which all common activities, erection and excavation, are to be measured. The two standards are naturally designed for use in contracts governed respectively by the ICE General Conditions of Contract and the JCT Standard Form of Building Contract. They are not, however, so restricted, and whichever standard is more appropriate to the work is selected: it is only necessary to make minor amendments to the conditions of contract to identify the version chosen and to ensure there is no incompatibility.

Bills of quantities are customarily prefaced by a *preamble* setting out the general methods and interpretations used by quantity surveyors in taking off quantities from the drawings. The same rules must be observed by the tenderer in pricing the bills and the Engineer in considering the tenders. The standard methods of measurement themselves recommend or indicate points which require attention in the preamble to the bills, in order to augment the definitions and rules which they themselves contain.

**4.18.3** Even in those countries (which include the UK) where it is usual for the Employer to embody ostensibly accurate designs, drawings and specifications (with full bills of quantities) in tender enquiries for constructional work, it is customary so to frame the contracts that they require the contractors to produce complete works even if the given quantities (against which they have tendered) prove to be inaccurate in practice.

In such countries, the relevant standard conditions of contract empower the Engineer to substitute, if necessary, during the progress of the contract, revised quantities based on a 're-measurement' of revised drawings or on the actual work necessarily performed. The contract price is also suitably amended in conformation. Such a practice is universal in building contracts in the UK, but for some types of civil engineering work it may, reasonably, not be adopted.

Whenever it is the Employer's intention to let a civil contract *without* remeasurement provisions, he must make this clear to tenderers in his enquiry, and delete the relevant clauses (permitting remeasurement) from the contract conditions he puts forward. Such a contract would, of course, still retain the normal contract variation procedures, so that any errors or omissions discovered subsequently could be rectified by this means.

In plant contracts, however, responsibility for detailed design customarily rests with the supplier, so that the need for remeasurement does not normally arise. Some provisions of the same nature are, however, often introduced into contracts for cabling or pipe-work where sizes, routes and lengths cannot be precisely established at the time tenders are called for, or are dependent on plant, roads or buildings not yet available.

If, at the time of going out to tender, the design of the works is insufficiently complete to enable meaningful quantities to be stated, bills of quantities are replaced by schedules of unit rates. It may be possible to include broad estimates of quantities as an indication of the magnitude of the works, but no attempt is made to evaluate items and arrive at a total tender price: competition is based on quoted rates.

**4.18.4**   The taking-off of quantities from the technical documentation and in particular the drawings is the responsibility of the quantity surveyors. Close liaison is necessary between them and the engineers responsible for the design and specification of the works to ensure the drawings from which the quantities are taken are the latest revision as destined for issue with the enquiry. Design work is seldom at a stage where it is completed and put down by the time the working drawings are sent to the quantity surveyors for preparation of the bills to start. Subsequent revisions must be fed to them as soon as they can be made available and 'last-minute' changes in the tender drawings must be notified as far ahead as possible. Otherwise discrepancies between the bills and the drawings will result.

Tenderers should always be asked, in the instructions to tenderers, to notify the Engineer immediately of any discrepancy they may notice when making up their prices. It can then be notified and rectified for all tenderers by a formal amendment (see Section 5.6).

**4.18.5**   So called 'bills of quantities' are sometimes met with in tenders for the supply and erection of electrical or mechanical plant under conditions of contract other than the ICE and JCT Standard Conditions. They are not, in fact, true bills of quantities of the sort we have been describing but are 'shopping lists' defining the scope of the apparatus to be supplied under the contract. Their main uses can include:

- Definition of scope of plant included in tender.
- Definition of 'Sections of the Works'.
- Breakdown of lump-sum tender price for use in connection with tender appraisal, split contracts, price levels for accessories, etc.
- Use during contract in connection with contract price adjustment, liquidated damages for delay in completion, interim payments on part deliveries.
- Establishing rates for supply of replacement parts, running spares, insurance purposes.

## 4.19   PRE-CONTRACT ESTIMATE OF THE COST OF THE CONTRACT

At about the time an enquiry is to be issued it is not unusual (especially with civil engineering or building contracts) for the Employer to call for a pre-contract estimate of the cost he is likely to face for the works. Even though some estimate may have been made previously as a part of his budget for the project, the whole concept of the project and the form of the contract works may have altered in the meantime, as may the level of prices in the industry.

Once the design for the contract works has crystallized sufficiently to allow an enquiry to be issued and has been expressed in realistic bills of quantities, a close estimate can be arrived at. The bills of quantities as drawn up for the enquiry documents are priced by the quantity surveyor using prices and rates current in the locality of the site. With prime cost sums based on the latest negotiations with intended nominated sub-contractors and provisional items on the most recent forecasts, a reasonably close estimate can be reached for a contract price.

It has three important functions:

- It enables the Employer to call for economies before any enquiry goes out, should he consider the present design of the works is cast in too ambitious a mould.
- It gives the Employer early warning of the likely cash-flow commitment he has to arrange to meet during the period of the contract.
- It gives the Engineer a useful yard-stick against which to compare the actual bids received from tenderers in response to the enquiry.

# 5 *The tender period*

## 5.1 GUIDING PRINCIPLES FOR THE ENGINEER DURING THE TENDER PERIOD

The *tender period* is the time between the issue of the enquiry to tenderers and the date specified for the return of the tenders themselves (including any formal extension of such date notified to tenderers subsequent to the issue of the enquiry).

In the interests of properly competitive tenders and by reason of his dual role (as consultant to the Employer and as mediator between Employer and contractor), the Engineer must be guided in his actions during the tender period by four basic principles:

- Strict impartiality must be exercised between tenderers even on matters of detail, and must be seen by them to be so exercised.
- Complete security must be maintained to prevent any relevant activity or proposal by one tenderer becoming known to another through the Engineer.
- Recognized rules and procedures must be carefully followed so that if either of the first two objectives is breached, it can be shown to be through no fault of the Engineer.
- Within the limits set by the first three principles, the Engineer's actions and decisions must reflect his duty to his client, the Employer, and the latter's best interests.

It is important for the Engineer to avoid among tenderers any justifiable sense of grievance on the grounds of having received less-favourable treatment than their competitors. Especially if the resulting contract is a valuable one, or one of particular importance to a tenderer, his resentment at being unsuccessful because of supposed partiality by the Engineer may persist for a long time, to the detriment of the Engineer's reputation and his future relations with the tenderer concerned. Even if the tenderer is successful, the Engineer's position may be prejudiced if he is called upon to mediate in his role as quasi-arbitrator, which we referred to in Chapter 1.

## 5.2 RESPONSIBILITY FOR ISSUE OF THE ENQUIRY

**5.2.1** The question of whether an enquiry is to be issued by the Engineer or by the Employer himself is a matter for early decision between them (see Section 2.1.2). If it is to be done by the Engineer, then the assembly of the master documents and the production and despatch of copies will usually be organized by his contracts engineer who will also ensure that:

(a) Instructions for the return of tenders to the Engineer have been included in the instructions to tenderers.
(b) An appropriate date and time by which tenderers must do this have been inserted before despatch of the enquiry.
(c) Official labels for return of tenders are sent.

The Employer must, in any case authorize the issue of the enquiry prior to documents being despatched.

If the documents are not too bulky to be returned in an envelope, suitable envelopes with return address labels affixed are usually sent with the enquiry, and the tenderer required to use none other when submitting his offer. Otherwise standard labels already completed (see Annexure to Appendix 8) must be included with each enquiry in sufficient quantity to enable the tenderer to return the documents expected from him in reasonably sized parcels. The aim of this procedure is to ensure that the cover used:

- Is correctly addressed.
- Is recognizable on receipt as a tender, to remain unopened until the named date.
- Identifies the enquiry.
- Identifies the contract engineer to receive it (unopened).
- Does *not* identify by name or any other mark the tenderer who sent it.

In certain circumstances there is some virtue in having tenders returned under double cover, the inner envelope carrying the standard label and the outer merely required to be clear of any identification of the sender. Whilst not normally necessary in tendering in the UK, the point should be considered when overseas countries are concerned. The outer cover can be addressed to an appropriate official of the Employer/Engineer by name. The procedure for dealing with tenders on receipt at the Engineer's office is of interest in the present connection (see Section 6.2).

**5.2.2**   The enquiry is sent simultaneously to all those on the final list of tenderers (Section 3.3) and a record of recipients must be kept. This is conveniently done on the 'Record of Enquiries Issued' pro forma suggested in Section 3.7.2 and Appendix 7. Either the letter of invitation to tender or the instructions to tenderers must call for immediate acknowledgement by the tenderer of his receipt of the enquiry documents, for which purpose a receipt form may be included with the letter. If the contracts engineer has not received confirmation of receipt from any addressee promptly, he should enquire by telephone whether the documents have in fact arrived.

The contracts engineer also deals with the internal distribution of the enquiry to the Employer (see Section 2.1.4) and the Engineer (notably the project manager, project engineer, the quantity surveyor – and the contracts engineer himself). The master copy is kept for the Master Contract Record (see Chapter 10).

Sufficient copies of the documents are sent to each tenderer to cover all those he is required to complete and return with his offer plus one complete set of all documents for his retention. It is left to the tenderer to prepare any parts he may need for enquiries to his intended sub-contractors. From the tenderer the Employer will require the return of:

(a) One complete set of *all* tender documents duly filled in, signed and dated in the appropriate places. They should be contained in one volume or one set of volumes, and they constitute the formal offer in all its aspects. A formal acceptance of the offer made by the Employer and the same documents automatically become the Employer's copy of the legal contract. One complete set of documents is therefore an essential minimum.

(b) The Employer/Engineer probably proposes to carry out tender appraisal simultaneously by two or three persons, for example, the project manager, the project engineer and the contracts engineer. For their use, the tenderer must be required to return extra copies of those parts of the tender documents which he has been required to fill in with data relevant to his offer, for example, two extra copies of the form of tender, pro formas for data, bills of quantities, schedule of points of non-compliance, etc.

### 5.3   LIAISON WITH TENDERERS

During the tender period all liaison with tenderers by the Engineer (and preferably the Employer also) should be channelled through one person (often the contracts engineer) who must be specified by name in the instructions to tenderers. Tenderers must also be advised that enquiries relating to the tender should be by letter or telex and not by visit or telephone. This is to avoid the difficulty during discussions of preventing disclosures of information denied to the other tenderers. Even if it does not happen, there remains the difficulty of proving it, should subsequently a leakage be shown to have occurred. We deal further with this matter in Section 5.4.

Other engineers, who may be approached by a tenderer, must refer him without comment to the appointed channel, even though they may themselves be better qualified to answer the tenderer's enquiry. The contracts engineer will obtain their views and transmit them to the tenderer without being tempted to elaborate on them.

All telephone conversations or meetings with a tenderer must be noted, preferably with a précis of what transpired and certainly with a statement of any decisions reached. The enquiry log we suggested the contracts engineer might keep (see Section 3.7.1) is well suited for this purpose, details being added as a separate sheet if necessary.

Any decisions made must be immediately confirmed in a letter to the tenderer and his agreement in writing obtained. This can conveniently be done by a suitable statement typed on a duplicate copy of the letter itself which the tenderer signs and returns.

## 5.4 DISCUSSIONS WITH TENDERERS

**5.4.1** As we have already indicated, discussions with tenderers must whenever possible be prevented, even in face of charges of non-cooperation. However, if verbal contact with a tenderer is unavoidable he should be at once invited to state his purpose in one or more direct questions which the contracts engineer can then:

- Refuse to answer.
- Defer reply for consideration.
- Answer specifically, without embellishment.

Conversations and 'chats' with tenderers are most undesirable as it is difficult to appreciate during them just what extra information is being given, implied or hinted at.

The contracts engineer must reject all queries which seek to give the tenderer an advantage over his competitors or information on their activities. As a guide, the contracts engineer should need to be convinced by the tenderer that an answer is really necessary, rather than having to convince himself he need not reply.

The deferred reply is useful in many instances, quite apart from those in which the contracts engineer does not know the answer himself. It enables the problem to be discussed, implications or policy matters arising to be referred for decision to the project manager or to the Employer, the checking of data, or it can serve purely as a means of avoiding being drawn into a discussion with the tenderer, if such appears imminent.

Questions involving only the interpretation of the tender documents, especially if they involve obvious ambiguities in the wording or unintentional omissions, can usually (after due consideration in the light of the foregoing) be answered direct. However, even in these cases, the answer will often be of interest to all tenderers and can best be made the subject of an official amendment to the enquiry (Appendix 21). If it has been decided that the question can suitably be answered direct, the tenderer will be told that the information will be confirmed in writing in the form of an amendment.

**5.4.2** In spite of what has just been said, there are certain types of enquiry (even involving competitive tendering) in which basic principles have to be ignored and matters discussed with tenderers. The proceedings will, however, usually be treated in close confidence.

One example is the general meeting of tenderers which forms an important part of the procedure in a well-recognized version of 'two-stage' tendering. In this version, a provisional enquiry is first issued to all potential tenderers, in which the project is fully described with as much technical and design data as can currently be stated. No bids are called for, but after a period for study of the enquiry, a general meeting of tenderers is called at which the project is discussed with the Employer and the Engineer. All its aspects technical, commercial and contractual, are open to discussion and can be raised by anyone present. No detailed agenda is issued.

If individual tenderers so request, private meetings are also held with them at which they may again raise any questions, including the acceptability of proprietary solutions, or other original ideas which may be raised in confidence.

From all these views and discussions, the Employer and the Engineer decide which ideas are desirable and these are embodied in a new set of specifications and enquiry documents. This second enquiry is issued to all those who took part in the earlier proceedings, and is now for normal competitive bids. The specifications are made wide enough to embrace any attractive proprietary ideas and solutions, and confidence is maintained by not mentioning the firms or giving sufficient detail in the specification to expose the method.

Such a procedure, which in effect picks the brains of all the most likely contenders, is most useful in projects which contain a large 'unknown' element.

**5.4.3** Returning to the original point, which was the extent to which enquiries could be discussed with individual tenderers in face of the principle of equal competition, it can be said that any tender in which design responsibility is placed on the tenderer can readily give rise to discussions with tenderers individually, during the tender period, on such matters as the technical acceptability of their proposals, preferences for modified features of design, relative priorities of conflicting requirements and such like. In these discussions a tenderer will probably disclose matters and designs which are novel or proprietary and clearly confidential. The Engineer must take the greatest care not to divulge or suggest them during subsequent discussions with other tenderers, even though their adoption might lead to a better solution for his client, the Employer. It could be unethical certainly, and possibly actionable.

The position becomes even more dangerous if the disclosure to the Engineer involves novel features which might rank as an invention, especially if a leakage does somehow occur: the Engineer will always be open to suspicion (and indeed accusation) of having been the source of the leak. It is a sound principle to refuse to listen to any disclosure by a tenderer of any novel feature until the tenderer has applied for and obtained provisional patent protection. The process is inexpensive and takes only a day or two, and the rights of the inventor are then protected against future infringement or unauthorized use. Confidentiality must still be strictly observed by the Engineer against any loss of competitive advantage, but the greater risk arising from unauthorized publication of an invention is overcome.

## 5.5  POSTPONEMENT OF THE DATE FOR THE RETURN OF TENDERS

Requests for postponement of the date given in the tender documents for the return of tenders are not granted just for the asking. Assuming the original date was one which experience indicated was reasonable for the production of a tender of such complexity by a competent firm, several points need consideration:

● Would a revised date involve a consequent delay to the project programme?

- Would postponement imply preferential treatment of one firm over its competitors, who can meet the original date?
- Does the enquirer produce any convincing reason why his tender is being unusually delayed? Or is it due to incompetence or inexperience in the work to which the tender relates?
- Would a refusal to grant an extension result in an insufficient number of truly competitive tenders being received?
- What are the Employer's views?
- Is the enquirer someone from whom a fully considered tender would be welcome? Has he perhaps appreciated difficulties in the works which others may have overlooked?

Any extension of the tender period granted to one tenderer must be notified promptly by telex or telephone to all other tenderers and confirmed to all by a formal amendment letter.

When any amendment of substance is made to the enquiry documents by the Engineer during the tender period, he must at the same time consider the amount of extra or repeat working which tenderers may be involved in thereby. Any appropriate extension of time for the return of tenders can then be authorized in the same letter as the amendment itself.

## 5.6  AMENDMENTS TO ENQUIRY DOCUMENTS BY THE ENGINEER

Any amendment to enquiry documents after issue will be formally notified to all concerned by an amendment letter issued by the contracts engineer (see Appendix 21 for a typical example). The letter must include a return slip (or a duplicate copy) which is used by the addressee to acknowledge its receipt. Amendments will be serially numbered. Their issue is recorded by the contracts engineer in his tender log and a copy of the amendment is kept for the master contract record (Chapter 10).

The same procedure is used to rectify errors or omissions discovered and notified by the Employer or tenderers themselves. Amendments must be succinct and clear, in a non-discursive form such as: delete '. . . . . . . . . . . .', substitute '. . . . . . . . . . . .'. Long revised passages can usefully be sent in a complete sheet for substitution in the enquiry documents. A tenderer should be sent a copy of any amendment sheet for each copy of the enquiry documents in his possession and be required to amend the latter before returning them as his tender. Either on the receipt form for the amendment or, preferably, with the form of tender, the tenderer must certify that his tender will take (or takes) full account of the amendment.

An amendment, as issued, must be complete in itself and not rely on the tenderer appreciating its implications and having to decide what other changes to the documents must be introduced in consequence of it. Thus, an amendment in respect of a minor change in specification must also detail the changes it implies in the drawings, bills of quantities, schedules of drawings, changes to test procedure, etc. The advantages of issuing replacement sheets which the tenderer merely substitutes for the original must always be borne in mind.

## 5.7  TENDERERS' VISITS TO SITE

The instructions to tenderers (Section 4.6) must lay down that no tenderer may visit the site without the prior authorization of the Engineer and after carrying out any other procedure which the Employer or Engineer prescribes.

A representative of the Engineer may meet the tenderer and accompany him during his visit to the site. He can then ensure that the information given to the tenderer is confined to pointing out on the site features which are referred to in the enquiry documents and the matters of fact concerning site conditions or topography. He can also take note first-hand of any queries raised by the tenderer which may call for a later reply in writing or the issue of an amendment to all tenderers. Especially where such queries indicate an inadequacy in the enquiry documents, the project manager will confirm with the Employer that the latter has no objection to additional data about the site being released, and if so the contracts engineer will incorporate it into a formal amendment to all tenderers. The tenderer visiting the site is meanwhile told that his request will be considered and if accepted, the information sought will be sent to him. The Engineer's representative must then ensure the question is followed up immediately.

The Employer may also require his representative to accompany the visiting tenderer whilst on site. If he does, the Engineer's representative must ensure, in the interests of impartiality, that no new data or additional site facilities over and above those notified in the enquiry documents are given to the visitor, or if they are so given, the same are offered to the other tenderers without delay, and in time to be taken into account in their offers.

## 5.8  REJECTED INVITATIONS TO TENDER

Any contractor who declines to tender will have his notice acknowledged, accepting his refusal, and will be asked to send back all enquiry documentation not already returned. Copies of the correspondence will normally be sent to the Employer and the circumstances recorded in the record of enquiries issued (Section 3.7.2). If not stated in the contractor's rejection notice, or if a contractor merely fails to submit a tender, the contracts engineer should endeavour to find out the reasons for the decision: they can often be sufficiently interesting to warrant inclusion in the contractor's file in the register of contractors (Section 2.3.1).

## 5.9   AMENDMENTS TO TENDERS BY TENDERERS PRIOR TO THE DATE APPOINTED FOR TENDER OPENING

Tenderers may, subsequent to submission of their tenders but prior to the time of formal opening of tenders, withdraw them completely by notice in writing to the contracts engineer (unless they have entered a formal agreement not to do so). The withdrawal notice will be retained in safe custody with the tenders received until formal opening, when it will be possible to identify the tender concerned and withdraw it from the competition. The notice of withdrawal will be acknowledged on receipt, but the tender when withdrawn is not normally returned to the tenderer.

Tenderers may submit amendments to their offers during the period prior to the time of formal opening of tenders. These amendments must be made in writing, enclosed in plain envelopes and addressed in exactly the same way as he tender *except* that the envelope or label will be identified as 'amendment to tender'. The amendment will be retained unopened with the tenders themselves until the formal opening time. Amendments, signed by an authorized official, are acceptable in the form of a simple declaration of the changes to be made to the original tender, or as a resubmission of parts or the whole of the tender, but each page of the amendment should be identified by the initials of the tenderer's authorized official before submission.

The above comments are made in respect of English law and are generally true both in the UK and in countries whose legal code closely follows that of the UK. They are not, however, true in a number of other countries overseas where offers, once submitted, have to be kept in force unchanged, and open to acceptance for a reasonable period, or for the validity period if one is stated.

Legally the remarks above are true over the whole period up to the acceptance or rejection of the tender or the expiry of its validity period (if this is sooner); the practical restrictions which have to be applied to changes *after* the tender opening date are different, for reasons which will become clear (see Sections 6.9 and 6.10).

# 6 Receipt and custody of tenders – tender opening

## 6.1 SECURITY OF DOCUMENTS

From the moment tenders arrive on the Engineer's premises their security must be tight, rigid and routine. Industrial espionage is not just a subject for television dramas – it exists. Rarely is it carried out by highly organized teams: much more usually it arises from a casual unauthorized scrutiny of a security document left lying about, unattended, on an office table. Normally it is not the work of a professional, but someone, unscrupulous, who has a 'friend' in the opposite camp who will 'see he is alright' for information supplied.

With large contracts at stake it can be worth a lot to a tenderer establishing his bid to know the attitudes, methods or bids of his competitors.

If any leakage occurs, the Engineer's reputation depends on his being able to show beyond doubt that, by virtue of his strict security measures, the leak could not have resulted from the actions or inactions of his staff.

Security should be made the definite responsibility of a named member of the Engineer's staff (e.g. contracts engineer) who will draw up and enforce a written routine for the safekeeping of documents.

## 6.2 RECEIPT OF TENDER DOCUMENTS

By warning the Engineer's post room of the impending arrival of the specially labelled parcels, the contracts engineer ensures that tenders, as soon as received, are placed in his hands unopened. To enable such parcels to be recognized they must be correctly labelled by the tenderer with the Engineer's standard tender label, and this entails appropriate instructions being issued with the enquiry (see Section 5.2.3). Tenders delivered by hand should be taken by the bearer to the post room who must have instructions to hand the bearer a receipt, dated and timed. The date and time of receipt must also be recorded on the tender cover.

The contracts engineer will examine packages for any signs of having been opened and reclosed, and when satisfied will give each a serial number which, with the date and time of receipt and his signature, will be written on the face of the package and an appropriate entry made on the contracts engineer's enquiry log (see Section 3.7.1).

Any signs of tampering must be investigated by the contracts engineer to try to establish how and when (and if possible, by whom), it was caused. In practice, of course, most cases occur either by error in the Engineer's post room or before despatch from the tenderer's office because some document had been omitted or some error was discovered and corrected. Such cause, however, must be confirmed, and the finding recorded in the contracts engineer's enquiry log.

The tender is then placed, together with any others received against the same enquiry, in a safe, strongbox or cabinet under lock and key there to remain until the date and time appointed for formal tender opening. The invariable use of the Engineer's standard tender label (Appendix 8), and its proper completion by the contracts engineer before despatch with the enquiry documents, helps greatly in identifying the contents, especially when several enquiries for the same project are out to tender at the same time.

## 6.3   TENDERS BY TELEX

Some 4 hours prior to the time appointed for tender opening, any firm on the tender list from whom it is suspected a tender might not have been received should be telephoned by the Engineer to ask if their tender has been despatched. If it has, the firm may be invited to submit its tender price by telex, and on receipt it will be held by the contracts engineer pending the arrival of the tender documents themselves. *Under no circumstances* must tender prices be accepted or discussed by telephone: such a proceeding is open to many forms of misunderstanding and malpractice. When the tender duly arrives the tender price is confirmed as being in agreement with the telex, and the date and time of posting are checked from the postmark. If confirmed, the tender is given full validity (provided, of course, that the telex quoting the tender price is in the Engineer's hands before the time appointed for tender opening).

Tenders brought by messenger may also be accepted late at the discretion of the Engineer provided that:

- The tender price has been notified by telex prior to the opening time for tenders, and agrees with the written document.

- There is adequate evidence that the messenger left in reasonable time but was delayed.
- The arrival of the messenger was not unreasonably long after the official opening time for tenders.

### 6.4 TENDERS ARRIVING LATE

**6.4.1** Unless the tender price has been established by telex prior to the appointed opening time, tenders arriving thereafter are not normally considered in the competition. They are endorsed 'Received Late' by the contracts engineer, marked with the time and date of actual receipt in the Engineer's post room, and signed and returned forthwith to the tenderer.

**6.4.2** Relaxation of the above rule may be allowed at the discretion of the project manager in circumstances such as the following:

- Where there is clear evidence (e.g. by postmark) that the tender was despatched in time to have arrived by the appointed date under normal circumstances but was unreasonably delayed in transit. This may frequently be the case with tenders from overseas which suffer customs delays or the like.
- Where only a single tender was invited.
- When no tenders were received in time. The project manager may then grant an extension of time to *all* tenderers.
- When in the opinion of the project manager there are both exceptional justifying circumstances for the late arrival *and* it is to the advantage of the Employer that the tender should be considered.

A full record of events and decisions should be made by the contracts engineer in his tender log.

### 6.5 TENDER OPENING PROCEDURE

**6.5.1** At the appointed time, tenders must be formally opened in the presence of at least three people:

- The project manager or his nominated representative.
- The contracts engineer.
- The Employer's representative.

If the Employer elects not to be represented, a third member of the Engineer's staff, not concerned directly with the project or the tender enquiry, should attend as witness in his place. Tenders for opening will be produced by the contracts engineer and examined as being intact by the other members of the panel. They will be opened in turn and all copies checked against each other, particularly as to the tender sums. It must be established that all the required documents are present and that all issued amendments have been incorporated and taken into consideration. The declaration of bona fide tender (if required) must be present. Any discrepancies will be recorded on the record of tenders received. Any notice from the tenderer of amendment or withdrawal made subsequent to the submission of his tender is checked as being signed by an authorized signatory, preferably the same person who signed the original tender, and attached to the latter.

As each contract is opened the contracts engineer records details on the record of tenders received (see Appendix 23) which, at the end of the proceedings, is signed by all those present. It is a formal document of considerable significance to the Employer, Engineer and tenderers alike and must be treated as such. One copy of the record is retained by the Employer's representative or, if he is not present, must be sent to the Employer under confidential cover without delay.

## 6.6  UNANSWERED ENQUIRIES

Contractors who elect not to tender in response to the enquiry are required by the instructions to tenderers to return all copies of the enquiry documents to the Engineer (Section 4.6.2). If they have not been received within a few days after the formal opening, the Engineer must take steps to recover them. We have already commented on the value of establishing, if possible, the reason for a contractor's failure to tender.

## 6.7  TENDERS RECEIVED AND OPENED BY THE EMPLOYER

In the event the Employer decides that tenders are to be returned for formal opening by himself at his premises, the foregoing sections will not apply directly. In such circumstances the project manager should endeavour to arrange that he himself (or his contracts engineer as his representative) is invited to be present at the formal opening. He will take with him duplicate copies of the record of tenders received, which he will complete as the tenders are opened. Both copies are signed by those present, unless the Employer has instituted a similar procedure of his own, in which case his equivalent form will be used instead.

The Engineer's representative should bring back with him one copy of the form for the master contract record and at least one copy of each tender for use by the Engineer's staff during appraisal of the tenders.

## 6.8   ERRORS FOUND BY THE ENGINEER

**6.8.1**   The original of all tenders is first checked arithmetically immediately after the tender opening is complete under arrangements made by the project manager. The check includes a full computation of all the bills of quantities. If errors in arithmetic or pricing are brought to light they will be notified to the tenderer immediately, offering him the alternatives cited in Sections 6.9.3 or 6.9.4 as applicable. Where it is suitable, a form for error notification, such as the example shown in Appendix 19, can be used.

**6.8.2**   Even though no actual error is apparent, should a tender appear so low that an error is suspected, the tenderer should be asked to confirm his price. If he duly confirms a price that is clearly uneconomic, there is a danger its acceptance would involve the Employer and/or the tenderer in difficulties during the performance of the contract. The tenderer might therefore be asked to explain how he can justify so low a price, but failing a satisfactory justification, the acceptance of such a tender should only be recommended to the Employer by the Engineer with caution and reserve. If the tenderer does not confirm the price but indicates an error in its make-up, the situation is the same as in Section 6.9.3 (or possibly 6.9.4) and the tenderer is offered the same alternatives.

## 6.9   AMENDMENTS TO TENDERS BY TENDERER AFTER THE DATE OF OPENING OF TENDERS

**6.9.1**   Between the time of opening of tenders and a contract being awarded, a tenderer may, on his own initiative, ask to amend his offer. If the amendment is such as does not affect the prices contained in the offer it can be taken note of by the Engineer and either accepted as offered, or rejected, or negotiated in the case of tenderers on the short list, in the normal way.

The Engineer must be alert to the possibility of the amendment being an attempt to enhance the value of a tender following a leakage of information to the tenderer as to what his competitors have done: this would be taken into account in deciding whether or not to accept the amendment, and if the tender should be an attractive one, whether its selection also is affected.

**6.9.2**  If the proposed amendment affects the tender sum and there is, in the opinion of the project manager, no clear evidence of an error having been made in the original tender:

- If the amendment *reduces* the price, any decision on the award of the contract will be made on the prices *originally* quoted, but if the contract is, on this basis, awarded to the amending tenderer, it will be awarded at the reduced price.
- If the amendment *increases* the price it will not be admitted: the tenderer has to be told he may let his offer stand at the original price or withdraw it.

**6.9.3**  In any case in which the project manager is satisfied a genuine and bona fide error (not being an arithmetical error) has been made and was not built-in to allow the tenderer flexibility of price, he may at his discretion, and if it would be in the best interests of the Employer, accept the correction of the error and allow the corrected price to stand for consideration.

**6.9.4**  In the case of a clear arithmetical error the tenderer may be given the alternative of

*either*     Withdrawing his tender completely.
*or*     (a)  In the case of a lump-sum tender, confirming the original lump sum.
         (b)  In the case of a tender with bills of quantities, confirming original quoted rates, and correcting any errors in the extension of items in the bills.

If a lump-sum tender also shows a breakdown of price and the tenderer elects to confirm the original lump-sum price, he must, at the same time, indicate the adjustments he wishes to be made to the break-down prices in order to achieve arithmetic balance: if this is not done the Engineer will assume a proportional all-round adjustment of the broken down prices has to be made. In the case of a 'bills of quantities' tender, if the tenderer elects to confirm the original rates, the arithmetic correction of the defective extensions (and any totals involving them) will be made by the Engineer.

**6.9.5**  A tender error notification form such as that shown at Appendix 19 and its corresponding error confirmation form shown at Appendix 20 can be usefully used in dealing with errors under this section.

## 6.10  ACTIONS ARISING FROM AMENDMENTS TO TENDERS

**6.10.1**  All amendments to tenders after opening must be just as formally documented as the tender itself. Decisions by the tenderer must be specific and in writing, signed by a person duly authorized to commit his organization. Preferably this should be the same person who authenticated the original tender.

**6.10.2**  The contracts engineer is responsible for notifying all members of the Employer/Engineer staffs holding copies of the tender of all amendments confirmed or decisions to withdraw that are made.

**6.10.3**  There are considerable advantages in rationalizing the correction of errors by the use of the forms recommended in Appendices 19 and 20 in all tenders arising throughout a project.

**6.10.4**  It should be borne in mind that errors in a tender may not always be due to the fault, negligence or intent of the tenderer: they can arise from ambiguities, discrepancies or lack of clarity in the enquiry or from insufficient time being allowed to the tenderer to produce a carefully checked tender.

Tenderers are therefore entitled to be given sympathetic treatment in the matter of errors, as far as is consistent with the avoidance of malpractice or unfair advantage over their competitors. At the same time slipshod tendering is not to be encouraged and the rules set out above are designed to meet these ends.

## 6.11  WITHDRAWAL OF TENDERS BY THE TENDERER

Under English Law (though not under some foreign legal codes) a tenderer may withdraw his tender at any time before its validity expires or it has been unconditionally accepted. Such action can often be most troublesome in the later stages of tender negotiations, and in some instances (e.g. a tender for highly specialized services or equipment) a last-minute withdrawal can disrupt the whole timetable for the Employer's project.

The same can apply in the case of a nominated sub-contractor prior to the conclusion of his sub-contract with the main contractor. (See Section 8.2.1.)

Where withdrawal could be critical in this way, the Employer should be advised to enter into a separate Agreement with the tenderer or to require from him a bond or guarantee that he will keep his tender open for acceptance throughout its stated period of validity (which should be chosen in the enquiry to be sufficiently long to allow the tendering procedure to be carried to a conclusion), and that he will accept any contract awarded to him during that period and based on the tender.

This can be achieved in one of several different ways depending on the circumstances:

- By a formal Agreement between the tenderer and the Employer in which, as consideration for maintaining his offer, the tenderer is either paid a sum of money or accepts, as a service of sufficient value to him, the granting by the Employer of the opportunity to quote against the enquiry concerned.
- By rewording the Form of Tender so as to constitute a deed incorporating an undertaking to keep the offer it contains open to acceptance by the Employer throughout the whole validity period. The Form of Tender must then be sealed with the tenderer's common seal and stamped with a 50p duty stamp, instead of the usual signature. See Section 9.2 below.
- By means of a guarantee by an acceptable third party as surety that in the event of premature withdrawal of the tenderer's offer or refusal to accept a contract awarded on the basis of the offer the surety will pay the Employer a stated sum of money. The amount is chosen to represent the extra cost to the Employer of the delay and trouble caused by re-tendering.

A further and rather less formal method is sometimes used, more often overseas than in the UK. A tender is only accepted for consideration against the deposit by the tenderer of a considerable sum of money which he forfeits if he withdraws his tender or if he fails to accept a contract awarded on it during the period of validity. This is known as a *tender deposit*, which is returnable if the successful tenderer keeps to his undertaking, or if the tender is rejected by the Employer.

### 6.12 DISTRIBUTION OF COPIES OF TENDERS

**6.12.1**  Rapid appraisal of tenders necessitates simultaneous work on the several aspects – commercial, contractual, financial and technical. For this purpose extra copies of tenders or selected parts of tenders may be required, the number depending on the urgency and the distribution of tasks to individuals. As befits confidential documents the number of copies must be kept to a minimum and they should be issued personally to the individuals needing to use them.

In many cases individuals require only part of the tender documents in order to carry out their appraisals and this serves to reduce substantially the amount of copying required.

**6.12.2** The responsibility for producing the required copies, issuing them and recording their recipients rests with the Engineer and is usually dealt with by his contracts engineer. Often sufficient spares are sent with the enquiry to the tenderer, who is required to complete them and return them with his offer.

**6.12.3** All copies (full or part) must be kept in a safe place when not in actual use, and not left lying unattended on desks or cabinets.

**6.12.4** There must be close liaison between the people involved simultaneously in an appraisal to ensure that any matters affecting one of them which appear in parts of the tender being examined by the others are brought to his attention. For example, differences in specification between two tenderers for the supply of plant noted by the technical examiner must also be brought immediately to the notice of the financial examiner. Preferably all differences should be expressed in money terms so that a fair 'value for money' comparison can be made.

**6.12.5** From many points of view (not the least being this matter of liaison) it is advantageous to allot a separate room with all appropriate security measures in which all concerned with a tender appraisal and assessment can work together. The larger and more complex the tender, the greater the advantages that accrue.

# 7  Tender appraisal

## 7.1  ORGANIZING THE APPRAISAL OF TENDERS

**7.1.1**  Tender appraisal is the examination and comparison of competitive offers received in response to an enquiry, and the recommendation (or selection) of the one most suited to the Employer's needs. If there is no clear choice, the appraisal decision may have to be deferred pending the acquisition of more data from, or the negotiation of unsatisfactory points with, some or all of the tenderers (see Chapter 8).

Each tender must be examined from three aspects:

*Technical appraisal:* by a competent project engineer, of the technical aspects of the offer and the constructional methods proposed.

*Contractual appraisal:* by a contracts engineer, of the terms and conditions to which the tenderer has made his offer subject, and any conflict they cause with those stipulated by the enquiry.

*Financial appraisal:* by a contracts engineer (or a financial specialist) of the build-up of the total effective price of the tender, the terms and methods of payment required by the tenderer, and the sums offered as liquidated damages.

**7.1.2**  An Engineer who has undertaken project management for an Employer must expect to carry out tender appraisal as part of his brief and be able to call on competent staff in each department to do it. An Employer may sometimes prefer to undertake tender appraisal himself and would then expect from the Engineer specialist assistance only, in one or more of the fields. Whichever way the appraisal is undertaken, the three aspects cannot be considered each in isolation: they are closely inter-related, an advantage in one department resulting all too often in a disadvantage in another. Technical excellence may be bought at higher prices, lower prices may depend on more generous terms of payment from the Employer or on more sensitive contract price adjustment factors, early delivery promises may be achieved by massive sub-contracting to small firms of unknown status and doubtful reliability.

The appraisal must therefore be a joint team operation with close liaison between participants and one member of the team appointed to act as organizer–co-ordinator and reporter. We have already mentioned the advantages of allotting a single room in which the parties making the appraisal all work. With large and involved tenders this becomes a virtual necessity.

**7.1.3** If the responsibility for making tender appraisals has been placed entirely on the Engineer by the Employer (as will be assumed hereafter) the appraisal team will be appointed by the project manager. The technical appraisal will be made by the project engineer or one of his staff specializing in the type of work involved in the tender. In practice it may often be the engineer who drew up the technical specification for the enquiry.

The contractual appraisal will be made by the contracts engineer allotted to the project concerned. He probably drew up the contractual aspects of the enquiry. Depending on the expertise available in his department, he or his colleagues may also cover some or all of the financial aspects of the tenders.

The financial appraisal covers a wide field, from the checking of tenderers' rates in the bills of quantities to the more esoteric effects of long-term buyers credit and the usages of discounting techniques (see Section 7.6). Depending on the Engineer's departmental organization it may involve contracts engineers, estimators, accountants, quantity surveyors, or an appropriate combination thereof. By the nature of his responsibilities generally for the contract, the contracts engineer will normally co-ordinate the financial appraisal.

The contracts engineer will usually be responsible also for the organization of the whole tender appraisal, assembling the findings of the different parties, making up the appraisal report and submitting it to the project manager for his approval before it is presented formally to the Employer.

**7.1.4** The timetable for production of an appraisal report will have been determined early in the enquiry as part of the enquiry programme (see Section 3.6). When issuing tenders for appraisal the contracts engineer must confirm or amend the date for resporting results in the light of the current project circumstances. The validity period for the tenders and the degree of likelihood that protracted negotiations with tenderers may be required must also be taken into consideration.

Arithmetic checking of tenders and their bills of quantities takes a lot of time and though it will have begun prior to the start of appraisal, it may not yet be complete. Tenderers' decisions regarding their alternative options for dealing with any errors may not have been received (see Section 6.9). A decision to withdraw a tender reduces the appraisal load, so that awaited decisions from tenderers should be sought as early as possible.

## 7.2    THE MAIN ASPECTS FOR CONSIDERATION DURING APPRAISAL

**7.2.1**    Whilst the work involved in making an appraisal will clearly vary with the size and complexity of the contract, the direction in which the bulk of the task lies usually can be foreseen:

(a) Fixed price tenders for building or civil engineering works to the Employer's design, not subject to remeasurement, will usually need attention to matters of non-conformity with the enquiry requirements. If these have been scheduled by the tenderer the load is reduced. Technical appraisal will be mostly concerned with methods of construction proposed, and their approval, and the tenderers' programme for execution of the works.

(b) Similar tenders subject to remeasurement introduce a further heavy investigation-load analysing the bills of quantities. In particular one is trying to isolate an overpricing of those items in the bills which might, in the normal run of events, be expected to increase on remeasurement. Underpricing elsewhere aims to restore the total tender price. Tenderer's views may also differ as to the ingredients they choose to include in the preliminary items of the bills as opposed to including them under items subject to remeasurement. Such variations can result in the Employer paying more for the contract. Locating the offending items in a tender is time-consuming and may require interviews with the tenderers concerned.

(c) In appraising tenders for plant against a functional specification, the bulk of the work will lie in the examination and analysis of the technical features of the competing designs, including their cost effectiveness. As was mentioned in Section 4.11, there is also a greater tendency with plant tenders to call for changes in conditions of contract, and negotiations on these and other matters are likely to be necessary with individual tenderers. Long-term credits, and the size and spacing of progress payments asked for, may also call for close appraisal. We deal with this in Section 7.6 and with negotiations in Chapter 8 below.

In the following sections are to be found representative check lists of the main points to be considered under the three appraisal headings – technical, contractual and financial. They are inevitably rather general since the variation in types of contract would make comprehensive lists never ending. A plant-supply contract has been assumed as probably embracing the widest range of points; it must be recognized that some of the points will not be applicable to the simpler construction-type contracts. The section dealing with financial appraisal has had to include a description of some of the more mathematical aspects of tender-price analysis, since an accurate comparison of tenders must employ them. In the end, if it can be fairly assumed that the enquiry was only sent to firms of status and repute who would be unlikely to produce a second-rate job technically, it is the financial picture which will usually be the crucial one in awarding a contract.

## 7.2.2 Technical appraisal

The main features for consideration must clearly vary with the nature of the works and the extent to which the tenderer is responsible for technical design. Where he has to meet a purely functional specification in the enquiry, the following aspects will be competitive and have to be considered in making an appraisal:

- Points of non-compliance with the specification and their effects on technical suitability.
- General suitability of tenderers proposals.
- Any items of unusual ingenuity or ways of easing the work of operation.
- Introduction of any extras, embellishments, or quality finishes not called for. Effect on quoted price. Could they be excluded if necessary?
- Data tenderers were required to provide on pro formae sent with the enquiry.
- Any checks possible on tenderers calculations, safety factors included, safety devices and cut-outs effective?
- Matters needing further clarification or explanation.
- Any alternative proposals made by tenderers, their technical or operational advantages/disadvantages.
- Construction and installation methods, suitability and strength of temporary works, space requirements, special handling equipment, safety considerations during site work.
- Tenderers proposed use and layout of site, effect on other contractors present, storage area requirements, any restrictions imposed on access, roads, or services.

- Weather protection for outdoor plant, painting specification (including surface preparation, shot blasting, galvanizing, etc.), treatment of assembly nuts and bolts, rivets, welds. Effective life guarantees.
- Tenderers proposed programme, including main stages, critical path, key dates, completion. Erection period.
- Proposed working hours, labour effort, overtime policy, contractors plant (quantity, suitability, types, sizes). Inefficient use of plant, excess waiting times.
- Details of critical items in tenderers designs, for example, in construction work: piling; joints in structural steelwork; mechanical drives; automatic shut-down procedures.
- Forecasts of running costs, operating costs, foreseeable replacement/ maintenance costs. Cost-effectiveness.
- Forecasts/guarantees on noise levels, vibration, exhaust pollution, effluents, temperature rises at full load, operation under overload.
- Acceptance and performance test specifications and standards which are offered/guaranteed.
- Terms of warranty on plant defects. Maintenance and emergency repair services available: speed of response to call-out.
- Scale and cost of recommended on-site spares for routine and running maintenance. Schedule of non-standard parts, long-delivery items.
- Documentation offered: scope, quantities and quality. Design drawings, cable runs, maintenance manuals. Any secondary masters supplied?
- Tenderers technical prowess, staff availability, reputation, similar contracts completed satisfactorily.
- Assessment of tenderers production loading, currently and during proposed contract period.
- Level of proposed sub-contracting, suitability of proposed sub-contractors. Choice of Employer's preferred sub-contractors.
- Training facilities for Employer's staff.
- Situation on licence agreements for patents used in the plant, terms of Employers sub-licence (to use and re-sell). Any royalties payable on production?

### 7.2.3   Contractual appraisal

It is here assumed that the enquiry, as issued, set out the proposed conditions of contract, site regulations, forms of tender as described in this volume. The appraisal will therefore comprise a check that the tenderer has properly completed the documents as instructed, together with a

consideration of all items of non-compliance and any additions, deletions or other amendments proposed in the offer (see Section 9.4).
  Check:

- Documents all signed in the required places by an official likely to be authorized to do so.
- Validity period of offer not shorter than requested.
- Outline programme of works included.
- Key dates proposed by Employer duly met? Magnitude of difference.
- Period of work on site? Testing of the works or sections of works properly organized? Criteria, recording and action if failure occurs.
- Terms of payment required, method of payment, currency, assumed rates of exchange.
- Rates of liquidated damages for lateness, or test failure: are they adequate?
- Contract Price Adjustment requirements included? Formula, indices, and base date stated?
- Warranty terms and period. Does it include diagnosis of fault and cost of substituting replacement parts?
- Schedule of sub-contractors supplied? Approved? Employer's preferred sub-contractors rejected, and if so are reasons acceptable?
- Nominated sub-contracting proposals accepted by tenderer? Proposals for meeting contractual problems accepted?
- Declaration of bona fide tender duly signed and returned?
- Site Regulations acknowledged and accepted in offer?
- List of contractual non-compliance items supplied? Implications of changes in conditions on working of contract. Acceptable to Employer?
- Any tender-letter sent with offer? Does it in any way qualify the offer or points of non-compliance?
- Any import/export authorities needed for the works, either in UK or in foreign country?
- Any licences, design rights, copyrights involved? Terms of licences and sub-licences. Rates of royalty, and who pays them? Any cost to Employer, especially on production rates?
- Tenderers proposals to meet legal requirements of any Acts affecting the works or operations on site, for example, pressure testing of boilers, pressure vessels; testing of cranes, cables, ropes; storage of inflammable liquids on site; noise levels; safety measures under the Health and Safety at Work etc. Act 1974.
- Insurance cover proffered. Terms of policy acceptable? Period of cover, risks covered, interests of Employer adequately dealt with?

Receipt for current premium. Loss and damage arising from theft, vandalism, strikes, strike-breaking and lockouts, civil disturbance all covered? (For foreign contracts see Section 9.4.7.)
- Are all bonds, guarantees or agreements called for by the enquiry properly supplied, or reasonable arrangements proposed, with suitable sureties?

### 7.2.4 Financial appraisal

The financial implications of the following points need appraisal. It will be noted that a number of the items have already appeared under earlier headings: in most cases there are two different aspects of the same item to be considered, and it is a salutary reminder how interlinked the three sections are – a change in one almost invariably has an important effect on another.

Certain factors in what follows are given further consideration in Sections 7.6 and 9.4.6 below:

- The tender price, is it subject to any discounts or reductions? Does it include any agency or other fees?
- Are any requirements of the specification classed as 'extras' to the tender price?
- Is the price arithmetically correct?
- Check the schedule of items of non-compliance. What is their effect on the tender price, i.e. cost of reincluding them?
- Does price include delivery, unloading into position, site testing, performance testing, proof-testing of vessels, cranes, ropes, etc?
- Does the defects warranty include cost of diagnosis of fault, delivery and installation of replacement parts?
- Is CPA applicable? Estimation of effect of likely inflation over the contract period on the cost to the Employer.
- Abnormal items in bills of quantities, schedules of rates, breakdowns of tender price, cost of extra items.
- Effect on tender price of any alternative schemes put up by tenderer for consideration. Cost-effectiveness of technical improvements resulting?
- Rates of liquidated damages offered, are they adequate? Are they related to anticipated damages Employer will incur?
- Terms of payment asked for, progress or interim payments, payment 'with order', buyers credit facilities (offered or asked, interest rate, periods for refund)? Minimum sum for inclusion in interim certificates?

- Calculate discounted prices. Net Present Value.
- Overseas contracts (the applicability of the following will depend on whether it is the Employer or Contractor which is in the foreign country). Are prices quoted ex-works, FOB, CIF or Delivered to Site? Who pays transit insurance, landing dues, stevedoring, import duties, land haulage, unloading at site? Who pays local taxes, consular fees, etc? Is payment of contract price subject to income tax in the foreign country: local value added tax? Terms of payment, currencies, methods. Letters of credit confirmed or not, method of calling off and where? Rates of exchange assumed in arriving at tender-price? International negotiability of local currency. Estimate of expenditure in local currency? Transfer of equivalent sterling authorized? Stability of local currency, forecast inflation rate, future fluctuations of exchange rates over period of contract. Status of tenderers local bankers? Do they have a London representative?
- Calculate cost to Employer of any licence fees, royalties, etc. which will be payable in addition to tender price.
- Estimated life of plant. Foreseeable annual operating and maintenance costs. Effect on NPV.
- Relative prices for replacement parts, and stock of spares at site.
- Percentage claimed by tenderer on prime cost items.
- Tenderers current financial status, capital commitments, cash/credit availability and sources.
- Financial status of sub-contractors proposed by tenderer.
- Tenderers reputation for prompt payment of sub-contractors.
- Tenderers reputation for making additional claims.
- If insurance policies referred to, is indemnity adequate? Is Employer included as principal?

### 7.3 THE TENDER APPRAISAL FORM

Major projects usually involve a considerable number of contracts for each of which a tender appraisal has to be made. The object is the same each time; namely to analyse and compare the tenders from all points of view and select the most suitable for recommendation to the Employer. To assist those making the appraisal to do so in a logical way leaving nothing of importance out, and to aid those having to read and digest the findings, there is much to be said for standardizing as far as possible the format of the appraisal report for use on all tenders. It becomes recognized and accepted: recipients know just where to find data they seek.

A suggested set of standard forms is given at Appendices 24 A–G. They are designed to strike a happy mean between being specific enough to make the information clear and readily available, and yet general enough to fit, flexibly, a wide range of types of contract. The set comprises four parts, some of which (Parts 1, 2 and 3A) will generally be required for all tender appraisals, whilst others (Parts 3B and 4) will only be used when the need arises: for example, large complex tenders, tenders for plant and machinery, etc.

*Part 1: The appraisal summary and recommendations.* This is the report proper. It summarizes for the Employer the details of the enquiry, schedules the tenders received, compares the highlights of preferred tenders and makes a recommendation as to the course the Employer should adopt. Part 1 is examined in more detail in Section 7.4, and the form is shown in Appendix 24A.

*Part 2: Summaries of each tender.* This part is an appendix to Part 1 in which the main features of each tender are extracted and analysed separately. Although the form is simple, it is an advantage to use it, so that a recognized sequence is used for each tender – normally one tender to a page. From the sheets, a short list of *preferred tenderers* (usually two to three) can be selected and compared in Part 1. Further details of Part 2 are given at Section 7.5 and a suggested form is shown in Appendix 24B.

*Part 3: The financial comparison.* This comprises two tables, supplemented if necessary by written analyses of each of the preferred tenders from their financial aspects.

*Part 3A: Tender-price comparison – preferred tenders.* A comparison on a single sheet, of the tender prices of three preferred tenderers, together with the prices of any alternatives and any options they contain. It also records adjustments made by the Engineer for appraisal purposes, to bring tenderers to a common basis of goods and materials. Part 3A can, if necessary, be supported by a written analysis of the contents of the form, comparing in more detail, the tender prices and terms of payment of the same three preferred tenderers. A sample table Part 3A is shown in Appendix 24C.

*Part 3B: Discounted prices and NPVs – preferred tenders.* This form is designed to assist the discounting of the proposed terms of payment and buyers' credits associated with a tender, to enable the preferred tenders to be directly compared on a common basis of their net present values. A separate sheet is required, of course, for each preferred tender. The form for Part 3B is shown in Appendix 24D, as a worked example to clarify the procedure.

(Further details of Parts 3A and 3B are given in Section 7.6. The findings from Part 3 can either be included in the individual tender analysis sheets of Part 2, or can be compared directly in the final report of Part 1, which ever is the more appropriate).

*Part 4: The technical comparisons.* Part 4 refers almost exclusively to tenders for plant or equipment, in which the tenderers each put forward individual designs, which have to be analysed and compared with each other and with the Employer's specification. The analysis is limited to preferred tenderers, and is made in three parts, each on a different form:

*Part 4A:* Comparison of plant and services offered; *Part 4B:* Comparison of performance of plant offered; *Part 4C:* Comparison of energy and supplies demands (compressed air, inert gases, water etc. as applicable).

The format of the tables must vary somewhat with the type of machinery being considered, but the principle and layout remain the same. The points of comparison of design, for example, will be different in almost all tenders, as will the measures of plant performance. Further information on Part 4 of the Form of Appraisal is given in Section 7.7, and typical forms for Parts 4A, 4B and 4C are shown in Appendices 24E, 24F and 24G respectively.

Modified tables can be used with some construction contracts for comparing specialized machinery-based methods of construction proposed by civil tenderers, as well as for machinery to be provided by their subcontractors.

## 7.4 TENDER APPRAISAL FORM – PART 1: THE SUMMARY AND REPORT

**7.4.1** Part 1 comprises the main report, the other three parts being in effect appendices containing the bulk of the contributory data. Of its three pages, pages 1 and 2 record facts relating to the enquiry (Paragraphs 1.2 and 1.4), the pre-contract estimate for the job (Paragraph 1.5), and schedules relating to tenders received (Paragraphs 1.6 to 1.8). Tender prices are recorded as quoted by the tenderer, including the correction of any errors which have been approved; they make no allowances or adjustments for differences between the technical and commercial proposals of the tenderers or for points of non-compliance with the enquiry.

**7.4.2** Paragraph 1.9 deals with preferred tenderers only, a short-list usually not exceeding three, which consideration of the tenders has indicated to be the most suitable offers. The section is a condensed summary of comparisons taken from Parts 2, 3 and 4 of the appraisal form. It features the highlights which have an important bearing on the ultimate choice of tenderer: the latter should develop logically from a consideration of Paragraph 1.9. The section must name the preferred tenderers and state for each one the corrected tender price, after adjustment in Part 3 of the appraisal form. If there is insufficient space on the form of Part 2, a separate sheet or sheets can be used, the form itself referring to them to avoid their being overlooked or misplaced.

**7.4.3** *The recommendation* to the Employer is given in Paragraph 1.10. It should always be a clean decision and recommendation of the course of action the Employer is advised to pursue. It allows no 'sitting on the fence' by the project manager. If possible the recommendation should:

- State the tender recommended for acceptance (and price).
- List any points of non-conformity with the enquiry which the Employer is recommended to accept.
- List any points which need to be negotiated before the recommended tender can be accepted, and the solutions the Employer should attempt to achieve.
- Nominate the second choice of tenderer (for consideration if negotiations with the first choice should be unsuccessful) and list the inferior features accepted by adopting it.

Exceptionally, if all the preferred tenders involve matters requiring negotiation so that a final choice cannot be made until the results of the negotiations are clear, the recommendation may be limited to arranging the preferred tenderers in a clear order of choice, on the assumption that all negotiations will be settled favourably to the Employer.

**7.4.4** The organization of an appraisal and the completion of the appraisal report are usually delegated by the project manager to his contracts engineer. As regard Part 1 of the form, the contract engineer's role is largely that of 'secretary to the committee' since the vital contents of Paragraphs 1.9 and 1.10 embrace all three aspects of the tenders and have probably been decided by the respective teams in discussions.

The selection of a recommended tenderer involves balancing the relative significance of views, usually conflicting, of the technical, contractual and financial specialists involved, all of whom inevitably attach prime importance to the pros and cons in their own field. At its simplest, the situation may be resolved by discussion between the teams during the course of the appraisal. The composition of the meetings and the degree of formality will depend on the complexity of the contract and the nature, degree and extent of the conflict of views. The responsibility for appreciating the need for a formal meeting and for arranging it then rests with the contracts engineer. In difficult cases, or if a conflict of views cannot be mutually resolved, the project manager must be brought in and he, with his closer knowledge of the Employer's intentions, will:

- *either* evaluate the points of conflict and select a recommended tenderer.
- *or* require the acquisition of further data from tenderers by enquiries or by negotiations.
- *or* refer back one or more points for further consideration by the individuals making the appraisal.
- *or* obtain further views/instructions from the Employer, especially on the relative importance he attaches to conflicting points.

**7.4.5** Whether or not the project manager has been thus involved in the actual selection, the recommended tenderer and the appraisal report itself must be approved and signed by him, and submitted by him to the Employer.

## 7.5 TENDER APPRAISAL FORM – PART 2: PRÉCIS OF TENDERS RECEIVED

**7.5.1** In Part 2 of the appraisal form, each tender is summarized and analysed on a separate sheet. Where a tenderer submits an alternative offer sufficiently different from the specification issued, a further sheet should be devoted to it.

The object is not merely to transfer data from the tender to a new sheet, but to extract those items in each tender which call for comment or yield deductions having a direct influence on the choice of preferred tenderers. There is no place for an item which is not followed by a comment or a deduction. Points of non-compliance of the tender with the enquiry requirements should be included, preferably as a schedule. Items not mentioned in Part 2 are assumed to comply exactly with the enquiry.

**7.5.2** Data may be given treatment in other parts of the appraisal form, and duplication should be avoided. For example:

- Financial aspects are given deeper consideration in Part 3.
- Plant and machinery specification, performance and power requirements are dealt with in Part 4.
- Comparisons between preferred tenderers are made in Part 1.

These should not also appear in Part 2 unless there is a valid deduction to be made from the data, not obvious in its treatment elsewhere.

**7.5.3** The different sections of Part 2 are completed by the appropriate member of the appraisal team. The co-ordination and assembly of the documents is usually the responsibility of the contracts engineer. The main subjects which need to be considered are shown in the check-lists in Sections 7.2.2, and 7.2.3, and 7.2.4.

**7.5.4** It assists routine completion of Part 2 if it is divided into standard sectional headings. This also helps to consolidate comments from a number of different people on any given subject. In practice it usually pays to align the headings with the major divisions of interest – technical, contractual and financial – but, in addition, to separate off the key subjects of price and time to completion. The example of Part 2 in Appendix 24B is shown divided in this way.

**7.5.5** The following should also be noted in the commentary:

● Absence or inadequacy of data provided by tenderer.
● The effect and importance of any non-compliance with the enquiry, and the effect produced on the tender price.
● With serious non-compliance, whether to reject the tender outright or to negotiate with tenderer.
● Any other matters requiring negotiation, and the desired objective when doing so.

### 7.6   TENDER APPRAISAL FORM – PART 3: FINANCIAL COMPARISON (INCLUDING DISCOUNTING PROCEDURES AND BUYERS' CREDIT CONSIDERATIONS)

**7.6.1** We have already spoken of the need to compare tender prices referred to a common basis of goods and services, with adjustments when tenderers differ in what they are offering for the money.

Under present-day conditions, and especially with contracts lasting several years, there is a similar requirement to compare tenders on a *common financial basis*, by making adjustments to reflect their different terms and conditions of payment and the differing contract periods over which payments are spread. The pro forma at Part 3B of the appraisal form is designed to assist this adjustment (see Section 7.6.6). The common basis selected is the equivalent total cost of a contract arising out of the tender expressed in terms of its present-day value, known as the *net present value* (NPV) of the tender. The adjustment is made by converting each payment the Employer is going to have to make in respect of the contract (as expressed in its terms and conditions of payment) into an equivalent sum paid (or set aside for payment) today. Between now and the actual date of payment such a sum could earn interest and increase in value depending on how long it is held awaiting payment. The more remote the date of payment the smaller its NPV. A tenderer requiring larger than average interim payments early in a contract will effectively cost the Employer more than another, quoting an identical tender price, who is satisfied with receiving payment later in the contract period. The greater the rate of interest, the larger the difference between them.

The process has to be applied to each and every payment the Employer is required to make under the terms of a contract arising from the tender. Although not directly involved in the basic process, such matters as contract price adjustment or fluctuations in foreign exchange rates may affect the actual sums the Employer has to pay, and if necessary, have to be allowed for in the manner explained in Sections 7.6.4 and 7.6.5.

The process of transferring back a future payment to its net present value (NPV) is known as *discounting* the payment. It is one of the normally recognized methods of applying the accounting technique of *discounted cash flow* (DCF). In parenthesis, the other methods are not usually applicable to tender appraisal problems, but are designed rather to assist decision-making within a firm's internal operations, such as a comparison of the return to be expected from alternative capital projects. Discounting is applied in such a case not only to payments made but also to receipts and recoveries expected to accrue, i.e. a true 'cash flow' year by year.

In our present task of appraisal of tenders, the discounting procedure usually tells us little or nothing of immediate interest about one tender in isolation: when comparing two or more tenders, however, it allows their respective total costs to the Employer to be compared directly, even though the amounts payable and the dates on which they have to be paid, are very different in each case.

In general, the results of discounting will not be significant in tenders less than £100 000 or with contract periods less than 12–18 months. Equally, of course, it will be superfluous in comparing tenders of the same price (or near so) if their terms of payment, times of payment and CPA formulae are all closely the same.

### 7.6.2 Buyers' credit

An Employer can rarely afford to keep sufficient uncommitted cash available for future capital purchases in the way inferred above. He finds it more advantageous to borrow money, on the best terms he can, to meet payments of the contract price as and when they fall due. Repayment of the loan, together with interest on it, and a number of other charges which are usually associated with it (such as commission, commitment fee, credit insurance, management charges, etc.), is made by the Employer in instalments spread over a period which often exceeds by some years the contract period itself. Often it is arranged that repayment shall not begin until the contract works have been installed and become operational, thereby earning revenue from which the loan repayments can be funded.

With home contracts the Employer is usually (but not always) left to make his own loan arrangements with his bank or with a merchant bank or finance institution which specializes in such matters. With contracts abroad, however, credit may often be available from government sources (or banking sources underwritten by a government) or from institutions interested in furthering world trade for one reason or another. International sources of credit of this sort include:

- Favourable credit arrangements made available by a government to home contractors, to be passed on when tendering to Employers abroad, as an inducement to place the order with them.
- A line of credit established by a government with the government of a foreign country with which it hopes to foster increased exports. Employers in the foreign country may be allowed to draw on the line of credit to fund contracts they place in the country which established it.
- In the case of large projects in approved countries, special credits may be given to selected Employers by the International Bank for Reconstruction and Development ('The World Bank'), or other international lending organizations (see Section 3.4.1).

Each source has its own rules and regulations, and usually restrictes to some extent the range of projects for which it will approve loans.

How are such loans, and the payments made by the Employer to redeem them, dealt with in evaluating competitive offers? If the loan and its terms of repayment form a part of a tender offer and become part of the terms and conditions of payment in an ensuing contract, then all such sums to be paid by the Employer have to be included in the discounting procedure. This would certainly apply, for example, to the first of the sources listed above, in which the buyers' credit is offered by the tenderer, and figures in the contractual terms of payment. This would still be the case even if the actual payment of the contract price to the contractor was made directly by the organization providing the buyers' credit, to be recovered from the Employer under the terms of the loan written into the contract.

If, on the other hand, the loan and its repayment do not figure in the tender they are a separate business transaction by the Employer, distinct from the tender being appraised. The Employer is deemed to pay those sums his contract (arising from the tender) requires him to pay, but how he arranges to find the money to do it, is his own affair.

Consequently repayments of this second type of loan do not figure in the discount process when making tender appraisals. (*NB* There are other operations inside the Employer's organization, customarily evaluated by a discounted cash-flow process, in which they do.) As a corollary to the foregoing it will be seen that buyers' credit which has to be included in the discounting process, frequently originates in a foreign country and is repayable by the Employer in a foreign currency: Appendix 24D is based on this factor.

### 7.6.3 The discount factor

The relationship between a future sum of money and its net present value (NPV) is referred to as the *discount factor*. It is arrived at by the well-known formula for compound interest calculations which states that a sum of £$P$ today attracting interest at $r\%$ per annum, compounded annually, will, in $n$ years time, become worth

$$£P \left(1 + \frac{r}{100}\right)^n.$$

Thus £1 today will in 4 years time at 10% per annum compound interest become

$$£1 \left(1 + \frac{10}{100}\right)^4 = £1.4641.$$

Looking at the calculation from the other end (i.e. the *dis*counting procedure) a sum of £1, 4 years from now, will (in our example) result from an investment today of

$$£\frac{1}{1.4641} = £0.6830.$$

The figure 0.6830 is the discount factor for 4 years and 10% per annum interest.

For daily use, tables of discount factors are available covering a range of years and interest rates, though the advent of the electronic pocket calculator has made direct calculation a matter of great simplicity. The NPV of any payment in a contract is the amount of that payment multiplied by its appropriate discount factor. If the rate of interest is expected to change over the period, the calculation of the discount factor can be made in steps, starting with £1 at the far end. If, for example, a present rate of 10% is expected to fall to 8% in 4 years time, the factor for 6 years is found by discounting 2 years at 8% followed by 4 years at 10%.

The rate of discount, $r\%$ per annum, is usually taken as the mean cost of capital to the Employer and as such will be decided by the Employer's accountants. An organization's funds usually come from a mixture of the following sources:

- Equities.
- Retained earnings.
- Long-term loans (including debentures).
- Short-term loans (including bank overdrafts).
- Depreciation.

These are subject to differing rates of interest, taxes and expenses which can be allowed for in calculating the mean cost of capital. For the purposes of tender appraisal (in which we are aiming to arrange tenders in an order of preference) it is unlikely that any small inaccuracy in the 'cost of capital' figure chosen will affect the *comparison* of tenders as long as the same figure is applied to all of them (though the *magnitudes* of the differences of their NPVs will be modified to some extent). Currently, figures of rate of interest 9–12% are common, with 10% as the most usual for organizations of limited risk.

Similarly, changes in discount factor which occur if compounding is applied more frequently than annually, are often of no significance but this can readily be checked if the choice of tenderer is found to rest on small differences. It should be noted that the concept of discounting derives entirely from the earning power of invested money. It has nothing to do with other possible changes in costs or value over the period such as might be caused, for example, by inflation. Some of these can be significant and must be investigated as separate issues, as explained below.

### 7.6.4   The effect of foreign exchange rates

The need for an Employer to make payments in respect of a contract in currency of some country other than his own usually arises from one of three causes:

- A tender is received from a 'foreign' country, requiring all, or some, of the payments to be made in the tenderer's home currency, or those of his sub-contractors.
- A tenderer is unhappy with the stability or convertibility of the Employer's home currency and elects to be paid in an internationally acceptable one.

- The tenderer, or one of his sub-contractors, has arranged buyers' credit terms backed by a foreign government which involve repayments by the Employer direct to the agents of that government in its own currency.

When comparing tenders in terms of their respective NPVs they are, of course, expressed in the Employer's normal (home) operating currency. But when a tender includes a payment which contractually has to be made in a foreign currency, the cost to the Employer which has to be discounted is how much it costs him (in his own currency) to buy the necessary foreign exchange, i.e. it must take into account the exchange rate between the currencies at the date the payment has to be made. Bids from tenderers in different foreign countries may each introduce a different exchange rate, each fluctuating in its own way.

The causes of long-term movements in currency exchange rates are legion and usually very deep-seated. They may well be due to influences outside the countries concerned: decisions by powerful financial interests (for example those investing the wealth acquired by the oil-producing states) can easily 'rock the boat' for the less-well-protected nations worldwide. It thus becomes extremely hard to forecast with any meaningful accuracy, the relationship between two currencies over periods of some years, unless they are formally or traditionally both linked together, or to a common base (such as gold or US dollar).

Comparison of tenders by their NPVs has considerable mathematical justification since the average cost of capital to a company (i.e. the discount factor) can be charted and predicted with some accuracy over the periods concerned. Unforeseen events will tend to affect all tenders in the same way and leave tender comparisons much the same. It would be a retrograde step, therefore, to adulterate the process by involving in it a factor so unpredictable as future international exchange rates. A procedure which has much more to support it is:

(a) To compare tenders by their discounted values (NPVs) in the usual way, using today's rates of exchange for converting foreign currency payments.
(b) To make a separate *qualitative* comparison of the relative susceptibility of each tender to exchange-rate fluctuations over the period of the contract it proposes.

In carrying out the latter, questions such as these may indicate the relative stability of tenders, or their tendency to move in favour of (or against) the Employer:

- What are the relative sums in the tenders being compared which have to be paid in foreign currencies?

- Are the currencies the same in all tenders?
- If different, what has been their relative stability history over the recent past?
- Are the currencies (or any of them) tied to gold, dollars, sterling?
- Are the currencies backed by a stable export in steady demand (e.g. oil, minerals) or are they influenced by unstable lines of exports (e.g. cocoa, tourism) only?
- Are currencies, on present trends, recognized as: (i) 'Hard' or 'soft' currencies? (ii) Liable to be devalued or re-valued upwards?
- What are the current inflation rates in the countries concerned, and are they rising/falling? Are the governments stable or shaky?
- Does the Employer have any regular exports to the countries concerned (the value of which would vary in the opposite direction to variations in contract price, and offset them)?
- Is a change in the Employer's own currency exchange status likely to affect all tenders equally or not?
- How susceptible are the foreign countries to changes in prices of basic world commodities, e.g. oil, wheat, rubber?
- To what extent (if any) can the Employer reasonably consider forward buying of currencies as an insurance against exchange fluctuations?

The answers to the two comparisons (a) and (b) above may well be such that a choice of a preferred tender can be made. It is only when the two answers balance one another that further investigations to establish a *quantitative* comparison in (b) are made, so that the actual cost of foreign exchange can be included in the NPV comparison itself. In such close tendering, the criterion for selection may well be technical or contractual superiority anyway.

Some indication of the relationship of two currencies can be ascertained by plotting their exchange rate over the economic storms and vicissitudes of the past few years. In this way some 'feel' may be achieved of their relative reactions to external world influences and the magnitude of changes that result. Alternatively some considered views may be available from a number of organizations who have an interest in future currency relationships, either as part of their business or of their general concern with foreign trade. Such might include:

- The international section of the Employer's bank.
- Financial organizations operating in the London Money Market.
- Information department of the British Overseas Trade Board.
- The Export Credits Guarantee Department.
- Association of British Chambers of Commerce.
- CBI Overseas Directorate.

In any case, it will be exceptional for data obtained to be other than tentative.

### 7.6.5 The effect of inflation on choice of tender

Inflation is the declining purchasing power of money over time. It shows itself during the course of a contract as a continuing rise in the cost to the contractor of executing the works. That is to say, compared with prices ruling at the date he made his tender bid, each item of work will cost him more pounds (though each pound will have a deflated value in real terms). Usually his tender arranges to obtain the extra pounds from the Employer by the inclusion in the contract of a price adjustment clause (CPA clause) tied to specified statistical factors which the contractor believes will adequately represent the additional pounds he requires on each payment. We examined the use of CPA clauses in engineering contracts in some details in Section 4.11.7.

The argument is sometimes heard that, by the date the Employer has to pay an increased number of pounds under the terms of payment, his whole operation will be geared to working in the deflated pound of the time, and it will therefore actually cost him no more, i.e. there is no need to take account of inflation when using discounting techniques of comparison. In some operations this can be true, but not when comparing the NPV of tenders. For example a company comparing the economic advantages to itself of two possible alternative capital projects will use a discounted cash-flow technique. For any future year of operation it will estimate the total cost of running each project and the additional income it expects to derive from operating them. If the company's products and capital turnover are such that it can rapidly adjust its pricing structure to match inflation, then the above argument is true; the revenue and outgoings in any year will tend largely to neutralize one another and no purpose is served by discounting them both.

In the appraisal of tenders we have a different situation. We are asking ourselves the question: 'What is the value *at today's prices* of all payments the Employer is going to have to make under the contract?' The NPV method of discounting has no built-in factor specifically to take care of inflation, though the ruling discount factor (which reflects the average cost of capital to the Employer) will certainly depend to some extent on any foreseeable inflation. Anyone lending the Employer capital today will naturally expect to receive a higher rate of interest if it seems likely that, when the time comes for his principal to be repaid, its purchasing power will have been eroded through inflation.

Basically the situation is no different from what it would have been if the tenderer had chanced his arm on the future rate of inflation, and quoted his estimated inflated tender price, with a schedule of correspondingly inflated payments in respect thereof, to be made on specified dates. That is to say, a fixed-price tender, without any CPA Clause. For tender comparisons one might estimate and substitute such inflated payments and discount them.

The problem in allowing for inflation in tender appraisal is the same as we found with foreign exchange in Section 7.6.4, namely, not whether it is necessary to take cognisance of inflation, but the great difficulty in estimating what allowances to make in terms of meaningful figures. The solution is, not surprisingly, also the same. Two separate appraisals of the tenders are made:

(a) A firm comparison of prices using the normal NPV technique and ignoring the affects of any CPA adjustments; and
(b) A qualitative assessment of the relative susceptibility of the tenders to increases as a result of inflation.

When the tenders relate to work all to be carried out in the UK and paid for in sterling, one can make the general deduction that in an inflationary climate, the contract having the biggest proportion of its payments payable furthest in the future will be most affected by CPA, and be most disadvantageous to the Employer. A weighted 'payment/due date' factor for each tender will be a reasonable guide to its susceptibility to inflation. And there is still the risk that the inflation rate may become negligibly small, or may reverse into deflation! Appraisals under (a) and (b) above may well, in many cases, enable a clear combined choice to be made; the decision becomes more difficult to assess when a tender preferred under (a) appears to be least attractive under (b).

When, however, foreign countries are involved in producing the whole or any parts of the contract works, decisions at once become much more tenuous. Each country will have a different potential inflation increase which one aims to assess. Even for a qualitative assessment one is faced with a series of questions very similar to (if, indeed, not identical with) those listed in the previous paragraph dealing with foreign exchange rates. Expressing feelings in reliable figures becomes well-nigh insuperable.

### 7.6.6  The appraisal form – Part 3

The tender financial appraisal is normally carried out on forms 3A and 3B (Appendices 24C and 24D), together with a written summary of those comparisons which cannot be quantified on the two forms. If the written summary is short, it can be included direct into Part 2 of the appraisal (at Paragaph 2.5): if more lengthy, it is set out as an addendum to form 3A and the conclusions only are noted in Part 2.

With tenders involving terms of payment or other financial matters of an unusual nature, additional forms can be added *ad hoc* to illustrate or compare these features.

*Part 3A: Comparison of tender prices*

This form records and compares on a single sheet, for three preferred tenderers, the breakdown of their tender prices and the cost of options or extras. An example of Part 3A is given in Appendix 24C. The subdivision of the tender in the left-hand column will, in practice, be prescribed in the case of lump-sum tenders by any break-down which the enquiry required the tenderers to make. In the case of constructional contracts with detailed bills of quantities, suitable subdivisions can be selected by logically grouping together items of the bills, it being a requirement, naturally, that every billed item must be included in one or other of the groups. Table 3A (Appendix 24C) then records (column 3) the amount of any adjustment made at appraisal to bring all preferred tenders to a common basis of goods and services, so that like is being compared with like. The particular feature in respect of which an adjustment is made is recorded on the reverse of the table or in the supporting commentary.

Financial considerations, which cannot be recorded quantitatively in the table of Part 3A, are recorded and discussed in a written financial commentary which compares such matters as:

- Relative forecast effects of CPA adjustments on the preferred tenders. Choice of base date and formula indices.
- Relative requirements for foreign exchange and the forecast effects of variations in exchange rates.
- Methods by which payments of the contract prices are to be made.
- Distribution of the works as between preliminary items and main items in the bills of quantities.
- Costs of spare parts and maintenance services.
- Comparisons of schedules of rates, or rates in the bills of quantities.
- Financial implications of proposed plant/labour ratios.
- Rates for liquidated damages, percentage on prime cost sums, etc.

*Part 3B: Discounting tender prices*

Form 3B assists the discounting of payments proposed in the preferred tenders, so as to yield in each case their net present value. It is shown, complete with a worked example, in Appendix 24D. In this form it is suitable for tenders with the contract price payable:

(a)  Partly in the currency of the Employer's country.
(b)  Partly in the currency of one or more foreign countries.
(c)  By one or more lines of buyer's credit, available and redeemable in a foreign currency.

If all foreign payments are to be made in the same currency, entries can be made in that currency and the total, after discounting, converted to the Employer's currency. When more than one currency is involved:

*either* a separate form is used for each currency, the total (after discounting) being converted to the home currency;
*or*    each payment is individually converted to the home currency before inclusion in the form, which is then discounted in home currency only.

The former method has the advantage of showing the NPV of each foreign currency separately: the latter only shows it for the total of all foreign currencies.

The calculation of discount factors for a number of odd periods is somewhat tedious, and it is customary to group together all payments occurring over a specified period, and discount them from the last day of the period. Six months is often chosen as a suitable period. For most contracts, with payments extending over 5 years or more, this approximation will make no sensible difference to the tender order. Of course, if contract payments have been fixed by time (as opposed to stages of progress of the works) and are payable at intervals of 6 months (or multiples thereof) discounting will automatically be made from the correct dates. Discount factors for intervals other than 6 months are not usually available from published tables of factors, but must be calculated as indicated in Section 7.6.3, using a pocket calculator.

The statement of buyer's credit shown in column 2 of form 3B (Appendix 24D) records the dates when sums are due to be drawn on the credit. They assist the calculations for columns (3), (4), and (5) but do not otherwise figure in the discounting calculations, as they do not represent payments against the contract price.

## 7.7  TENDER APPRAISAL FORM – PART 4: TECHNICAL COMPARISONS

**7.7.1**  The set of forms in Part 4 allows the Engineer to compare at a glance the plant for which tenderers have been responsible. The designs put forward by three preferred tenderers are analysed alongside each other on the same sheet. There are three forms, each dealing with a different aspect of the plant or machinery proposed (Appendices 24E, 24F and 24G):

*Part 4A:* Comparison of plant designs proposed by the three preferred tenderers. Connected services are also compared.

*Part 4B:* Comparison of the performance expected (or guaranteed) by the preferred tenderers from their proposed plants.

*Part 4C:* Comparison of the supplies required by the three proposed plants during operation, for example electricity, gas, oil, water, compressed gases or air, etc. These comparisons are usually made at full designed output load, but if light running is expected over long periods, comparisons at other loads may also be extracted.

**7.7.2** The main use of the forms in Part 4 is clearly in connection with tenders for the supply of plant and machinery, but the same principle may be useful in connection with constructional works, both in regard to any plant and machinery to be designed and provided by a sub-contractor and also in connection with any specialized, machine-based, method of construction which the civil engineering contractors have to devise and propose.

Highlights of the design of any tender are extracted from the forms of Part 4 to augment the data on the tender being assembled in Part 2 of the appraisal form, especially noting matters of operational efficiency. Comparisons can be extracted, if important, direct to Part 1.

**7.7.3** The schedule of items in the left-hand columns of each of the three forms are all special to the tender enquiry under consideration. For each enquiry they have to be selected by the engineer making the technical appraisal, to compare the key features of the machinery and its operation. In some cases the enquiry will have called specifically for data on certain points (see Section 4.9); in others the information will have to be found from an examination of technical documentation provided. Further data on key items may have to be sought during the appraisal periods by questionnaires to the preferred tenderers. The tables should also include items deducible by the engineer making the appraisal from data provided by the tenderer, for example energy consumption per tonne of product, floor space per tonne of product per hour, operating hours per year to meet annual production target.

**7.7.4**  When comparisons of machinery or its operation are transferred from the tables of Part 4 to Part 1 to be used in making a final selection of tender, they should be expressed, wherever possible, in terms of money. The final decision may be made by persons to whom the relative importance of abstruse technical comparisons may not be clear. For example, a statement of relative thermal efficiencies of two plants may not be immediately significant, but its translation into increased cost of a year's production will be readily appreciated and given its due weight.

# 8 *Tender negotiation, acceptance and rejection*

## 8.1 NEGOTIATIONS WITH PREFERRED TENDERERS

### 8.1.1 Prelude to negotiations

In his appraisal report to the Employer, the Engineer should nominate the tender he recommends for acceptance, and at least one preferred tender as 'runner-up'. (If more than one is recommended, they should be given an order of preference.) Tenders may not be fully informative, or completely in line with the enquiry issued: the appraisal report must therefore include a statement of any further data to be sought from, or points to be confirmed or reconciled with, each preferred tenderer. Matters of this sort are more likely to occur with tenders for the supply of equipment in which design responsibility rests with the tenderer, but are by no means so restricted. It will be normal for the clearing up of these questions to proceed concurrently with at least the first two preferred tenderers but negotiations must be kept entirely separate, both in regard to prices and to subject matter. They are for collecting information, not an opportunity to argue price reductions.

### 8.1.2 Negotiating procedure

Face-to-face negotiations across a table should be avoided whenever possible:

- They are time-consuming when time can ill be spared, owing to tender validities running out.
- It is difficult to prevent discussion leaving the point at issue and ranging wide.
- Tenderers, realizing they are on a short list, may wish to improve their price, and other aspects of their offer.

As a first step, whenever possible, points requiring clarification should be sent as a questionnaire, preferably calling for a 'Yes' or 'No' answer or the quote of a figure. More intractable points can often be settled by telex: 'Please telex by return the basis for your price calculations for. . . .' If, as a last resort, a meeting cannot be avoided it should be held on neutral ground (such as the Engineer's premises) and be conducted by the Engineer who must observe the following basic rules quite firmly:

- Negotiation is not horse-trading. Neither is it an opportunity for Dutch-auction between the tenderers.
- Discussion must be confined to specific questions, which must appear on an agenda.
- Discussion must never be permitted to become general about a tender, its merits or its chances of success.
- In no circumstances may the tenderer voluntarily offer or attempt to amend his tender-price, or other competitive aspects of his offer.

It may be no easy matter to achieve this objective in practice, especially with some overseas tenderers. Never enter the slippery slope away from the strict rules.


### 8.1.3   The negotiating team

For each question to be negotiated the team must include:

- *A spokesman:* who states the question and handles the whole argument of the Employer's/Engineer's case. He should be a member of the Engineer's staff.
- *A team-leader:* an Engineer's representative who stays aloof from the arguments, senses the way negotiations are going, leads the discussion in favourable directions by short interjections, and has the sole right to make concessions on behalf of the Employer as and when he feels they are necessary.The 'strategist'!
- *The Employer's representative:* a senior official of the Employer, with authority to permit concessions by the team-leader which involve extra cost to the Employer.
- *Specialist advisers:* as the subject matter may indicate, to advise the spokesman and the team-leader. They may be Engineer's staff or Employer's staff or both.

As an example, when negotiating a matter of basic plant design in a supply contract the team might be:

- Spokesman: Engineer's project engineer.
- Team-leader: Engineer's project manager.

- Employer's representative: Employer's project technical manager.
- Advisers: Employer's plant engineer, Engineer's contracts engineer.

The presence of the contracts engineer or his representative at all negotiations is important:

- To advise on the contractual implications of any changes to the tender which the negotiations introduce.
- To draft for the approval of both sides, any contract amendments needed to introduce a change.
- To ensure that any matters agreed by the meeting are properly recorded and introduced into the tender documentation, and are otherwise consistent with the terms and conditions of the tender.
- To act as secretary and to issue minutes of the meetings to all concerned (see Section 8.1.4).

The extent to which a position may be yielded by the team-leader in order to prevent deadlock must rest with the Employer who will be the ultimate sufferer: the team-leader should, if necessary, retire from the meeting with the Employer's representative to get approval of any unanticipated proposal the leader wishes to make. Much more preferable is to decide before the meeting starts, how far the leader may yield.

In the event the Employer elects not to take part in negotiations, the Engineer *must* agree with the Employer beforehand how far he may yield their position at his own discretion before having to refer back to the Employer. Negotiation between two parties is impossible if one has no freedom to move, and the need to break off discussions must be avoided. His freedom may be expressed in terms of resulting increase in tender price, or of technical limits or of arrangements specified as 'unacceptable', or a combination of these, and should be confirmed in writing before negotiations begin. In some cases the minimum concession the team-leader must obtain from the other side in exchange, the *quid pro quo*, may also be determined.

Meetings for negotiations (as opposed to meetings seeking interpretation or clarification) are seldom a matter of all 'give' or all 'take' but involve both sides yielding some concession the other wants. Questions can rarely be isolated from the rest of the tender but nevertheless the Engineer must resist the ever-present tendency to bring more and more of the tender 'into the melting-pot' in effect leading to a negotiated tender with one of the competitive tenderers.

### 8.1.4 Important aspects of negotiating technique

As has been said elsewhere, good negotiation is mostly an art combined with a flair, and is therefore not subject to definition by rules. There are, however, certain techniques and procedures which can smooth the way toward reaching a satisfactory conclusion, of which the following are the more important:

1. *Preparation is essential.* The negotiating team must, prior to the meeting:

    - List in writing the *precise* points which are in dispute. It crystallizes thought and serves as an agenda.

    (Opinions differ as to whether this agenda should be sent to the tenderer in advance to enable him to come to the meeting prepared, or whether in the interests of suspense and surprise, the items are divulged one by one at the meeting. In tender negotiations the balance seems to be firmly in favour of the former):

    - Obtain from the Engineer, in respect of each item raised: (a) the solution he wishes the team to achieve; (b) whether they can concede the point, in whole or in part, if pressed hard enough? (c) can they concede if amendments can be agreed elsewhere in the contract terms? (d) what specified alternative, if any, could be offered if the first fails? (e) is he prepared not to concede under any circumstances, and accept deadlock?
    - For each item in dispute, assemble their arguments in favour of their case. Stick to facts wherever possible: matters of opinion can always be disputed, but matters of fact are difficult to contradict.
    - Prepare 'exhibit' documents or copies for distribution. Locate papers on files and tag them for instant production to the meeting.
    - Consider tenderer's probable line of argument and prepare counters to it. If points needing a decision from higher authority can be foreseen, get this in advance or ensure Employer's representative present will be empowered to make it.

2. *The team.* This must be as small as possible consistent with pursuing the argument without having to adjourn for advice or approval. That shown in Section 8.1.3 is generally suitable, but overseas, specialist 'reserves' may have to be added:

    - Avoid being 'blinded by science'. If the tenderer decides to field an expert, the Engineer must have available someone equally competent to discuss the technicalities with authority, either present at the meeting or standing by 'outside' on immediate call.
    - Avoid being 'dominated by rank'. The status of members of the tenderer's team must be reasonably matched by that of the Engineer's, particularly in regard to the team-leader.

- The team can have only one leader who (in meetings at the Engineer's premises at least) acts as meeting chairman. There are advantages in the leader not being also the principal spokesman. The latter fights all the way for the chosen solution: the leader can make diplomatic moves or concessions to keep the momentum of the negotiation going without the spokesman having to withdraw anything he has said and lose face.

3. *The meeting:*

- Keep steadfastly to the matters on the agenda: complete the negotiations on each item before moving on to the next.
- The team spokesman for each item must have been nominated beforehand. The argument will, in general, be carried out through him. Supporters pass their comments or reminders to him wherever possible. Avoid a 'chatter of experts' all throwing comments at the tenderer's spokesman – he only becomes bewildered.
- The team must on no account argue among themselves in the presence of the tenderer's party. If a team-discussion is necessary, the leader must adjourn the meeting for a short recession.
- If negotiation loses momentum, avoid going round and round the same arguments. The chairman (leader) must either find a completely new approach (and it is here the true negotiator can really show his worth) or he must defer discussion, record 'no agreement', and move on to the next item.
- The meeting must be minuted, at least as to statements of principle made and decisions reached. The wording of any decision reached after much discussion should be read out and confirmed by the meeting before moving on to the next item.
- Avoid leaving loose ends, e.g. if the substance of a new clause has been agreed the actual clause must be drafted and presented to the meeting for approval before it ends.

A record of the meeting (the minutes referred to in 3 above) must be approved in writing by the parties present and circulated to those concerned (see Section 2.1.4). A copy should be retained by the contracts engineer for eventual incorporation in a master contract record, if the negotiations should result in a contract being placed.

It should be remembered in any negotiation that the tenderer also has his legitimate interests. By virtue of his responsibilities the Engineer must be fair to both parties. Any attempt to 'steam-roller' the tenderer into accepting something unreasonable, unnecessary, or unsupportable will only induce a more stubborn resistance to compromise, and gets the

Engineer branded as 'the Employer's hatchet man'. It may also influence his authority with the contractor during the execution of an ensuing contract.

### 8.1.5   The validity period of the tender

By this stage of the tender procedure, the validity periods of tenders will normally be coming to an end: if they are allowed to expire the whole enquiry may have to be retendered and reappraised. It may be necessary to negotiate with preferred tenderers an extension of validity sufficient to bridge the gap until all points can be settled and a contract awarded. Although for short periods most tenderers are prepared to agree to such an extension without change of price, for longer periods an inflationary business climate may make them decide that increases are necessary.

It is not unknown for a tenderer who sees he is high in the order of preference to be obstinate in negotiation from the start in order to force the Employer's hand – to give way or to accept an increase in price on revalidation. A final choice of tenderer may have to be considered as soon as agreement has been reached on the main points of issue, leaving minor ones to be negotiated after the tender has been accepted. Careful judgment of the contractual implications of failing subsequently to reach agreement, including the likelihood of such minor points becoming real issues during the execution of the contract, must be made by the Employer with the advice of the Engineer.

If, in the event, such a matter did arise and could not be settled amicably *ad hoc*, it would have to be referred to arbitration or the courts, and the cost might well exceed that of revalidating the tender and continuing negotiations. The use of an instruction to proceed (Section 8.5.1) should be considered as a possible solution.

### 8.2   NEGOTIATIONS IN CONNECTION WITH NOMINATED SUB-CONTRACTORS

**8.2.1**   During the period prior to placing a main contract, the Employer/ Engineer may be setting up a tender situation with any sub-contractor they intend to nominate to the main contractor and with whom the latter will be required, after appointment, to enter a nominated sub-contract. The reasons are mentioned briefly in Section 4.16.1.

To impose this method of operation on the contractor, it must be specified in the main contract, where 'nominated sub-contractor' must be defined and the working procedure and safeguards for the main contractor specified. This is done, for example, in the ICE Conditions of Contract (5th edition) and the JCT Standard Form of Building Contract, so that in practice the nominated sub-contractor concept is largely confined to constructional contracts based on these two standard forms of conditions.

Contractually, the operation can be quite complex, requiring expert handling and not a little 'tightrope walking' if it is to be flawless: a brief analysis of the situation in the first paragraph above, for example, will expose a number of potential dangers some of which were mentioned in Section 4.8.1 when dealing with the Form of Tender (qv). The following objectives need to be attained contractually:

(a) The main contractor must, after his appointment, enter the sub-contract; not the Employer. In doing so he must have some freedom to negotiate details.

(b) Nevertheless, prior to appointing the main contractor, the Employer negotiates with the sub-contractor he intends to nominate, what work he will undertake, at what price, and (at any rate in broad terms) the conditions of contract he will have to agree *vis-à-vis* the main contractor.

(c) Once (b) has been settled, the Employer wants the sub-contractor to stand by his offer and, later on, to enter into a sub-contract with some (as yet) unappointed main contractor. The Employer may well need some interim contract or bond so to bind the sub-contractor, rather than risk reliance on his good faith. He may himself enter a direct contract with the sub-contractor with the latter's agreement to its being assigned to the main contractor later (see, for example, the Form of Tender in Appendix 10C).

(d) Having set up this position, the Employer may still be thwarted by a valid objection by the selected main contractor to the choice of nominated sub-contractor he is expected to employ, or the scope of the work he is expected to sub-contract.

**8.2.2** In negotiating his nominated sub-contract, the main contractor naturally wants to pass on to the sub-contractor the identical terms and conditions of contract he himself has undertaken in his contract with the Employer. He is supported in this attitude by the standard forms of conditions of contract already mentioned, in which it is regarded as a principle that the same conditions must apply. This at once leads to anomalies, especially when the sub-contract is in a different discipline from the main contract, e.g. a main contract for a new roadway under ICE conditions (5th edition) with a nominated sub-contract for the provision of one set of automatic traffic signals, or a contract to build a factory with a nominated sub-contract for the elevators. We saw, in the Introduction to this book, how in each of these examples, the two disciplines operate in quite different ways, which are reflected in their own usual standard conditions of contract. We can rewrite the table of comparison we gave earlier, to contrast the main contractor and a sub-contractor in a different field:

*Main contract (constructional works)*
1. Contract direct with the Employer.
2. Design and drawings by the Employer. No contractor design responsibility.
3. Work almost entirely on Employer's site.
4. Price determined by remeasurement.
5. Rates shown in bills of quantities.
6. Payments usually monthly.
7. No site tests. Contractor only 'post-man' in plant operational tests.
8. No documentation.
9. No stocks of replacement parts.

*Nominated sub-contract (mechanical/electrical works)*
1. No contract with Employer.
2. Machine design and drawings by sub-contractor.
3. Work largely in sub-contractors factory.
4. No remeasurement. Contract price firm.
5. No bills of quantities. Lump-sum price normal.
6. Payments in two or three stage payments.
7. Acceptance tests and performance tests vital to Employer.
8. Full documentation (including maintenance books) required by Employer.
9. Must provide replacement parts over expected life of plant.

The use of nominated sub-contracting in such cases frequently necessitates amendments and additions to the main contract conditions to deal with at any rate the more important aspects of the other (sub-contractor's) discipline and its normal methods of working. There are also more fundamental problems that arise. For example the main contractor has no responsibility for the design of the permanent works under ICE General Conditions (5th edition), but he acquires in practice such a responsibility through his sub-contract, and must be made contractually accountable to the Employer. We revert to this subject later. On the other side, for example, the IMechE/IEE conditions do not recognize the nominated sub-contract system (though they do mention prime cost items) and suitable clauses introducing, defining, authorizing the system, its *modus operandi* and safeguards have to be introduced. However, the two standard conditions, each designed for its own kind of work, are so differently orientated that even quite extensive amendment does not make either a completely suitable substitute for the other. The relationship between Employer and contractor on the one hand, and the contractor-nominated sub-contractor on the other, can never, in such a case, be a very satisfactory or happy one.

**8.2.3** The nominated sub-contractor (usually with a contract only a fraction of the value of the main contract) often finds the main contract conditions he is expected to accept back-to-back unacceptably harsh. As one whom the Employer is trying to employ, he cannot see why he should be expected to abide by terms and conditions quite inappropriate to his type of work, and loaded with liabilities out of all proportion to the value of his sub-contract. His anxiety may be illustrated by four aspects of the conditions, as these typify the problems that can arise when the Employer is negotiating a tender with an intended nominated sub-contractor for the supply of plant.

*(a) The point of take-over and the start of the warranty period*

A plant contractor normally expects to perform tests on completion as soon as his plant is ready, and the period of warranty to begin as soon as the tests are passed. He is usually paid a substantial part of the contract price on passing the tests, and any outstanding balance at the end of the fixed-term warranty period. On passing the tests he becomes no longer at risk in regard to loss, damage or deterioration of the plant.

However, as a nominated sub-contractor he is expected to defer tests on completion until the main contractor can complete the works, test them as a combined plant, and hand everything over to the Employer together. The main contractor clearly does not wish to take over the plant himself (as the other party to the sub-contract) and retain the risk in it until the main works are finished. Neither is the Employer anxious to release the sub-contractor, with whom he has no contract, by accepting the ownership of his plant and the start of its warranty, whilst the main contractor is still at work on the project, and the plant cannot be used. As a result, the plant manufacturer remains liable for keeping the plant in tip-top condition and probably waiting for his money for considerable periods after he has left the plant complete and working on site. The start of his warranty period vanishes into the future when the Employer shall have taken over the whole main contract as a complete project.

## (b) Performance tests

It may frequently not be possible to carry out the prescribed tests until the whole of the rest of the main contract has been completed and set to work as a connected process. In the meantime (which can be considerable) the nominated sub-contractor must, at his own cost, ensure his plant is maintained at its peak performance (to avoid penalties later if it no longer meets the test specification). Any payments or retention money held by the Employer against satisfactory performance are withheld from the sub-contractor. He also remains at risk for an unspecified period outside his control for the penalities for eventually failing to meet his warranted performance.

## (c) Terms of payment

The main contractor usually wishes to withhold from the sub-contractor any payment he has not himself received from the Employer. This is no great problem when the main contractor's terms of payment from the Employer enable him to make an immediate claim for the amount concerned as an interim payment (though even here the delay in the pure mechanism of obtaining and transferring payment to the sub-contractor may be serious). When, however, the main contract terms do not allow the contractor to claim in his monthly statement of measured work, large lump sums of 'unmeasured' plant cost until such time as the plant work has been completed and tested as part of the finished project, there is a lengthy delay in the sub-contractor getting his money such as to become quite unacceptable to him.

## (d) Liquidated damages for delay

There are two separate points here, the first we deal with below and the second is set out in Section 8.2.5 (a). The main contractor, delayed in completing his contract works, may become liable to pay liquidated damages to the Employer at a rate related to the whole contract price. If the cause of his delay is delay by his sub-contractor, a problem can arise, and in particular when the main contractor is prevented from finishing off his own work by incomplete sub-contract work. Working on a back-to-back basis, the main contractor seeks to recover from his sub-contractor the full amount of damages that become due to the Employer, even though the value of the sub-contract may be miniscule compared with the main contract price on which the damages are calculated. The penalty is probably out of all proportion to the financial status of the sub-contractors business, and somewhat naturally, he objects. He cannot gamble his whole financial position on a single 'throw'.

The situation can be eased if the sub-contract works are self-contained and do not to any appreciable extent, impede the main contractor's own work. For example, the installation of lifts in a building, if delayed, should not impede the contractor erecting the building itself. The sub-contract works can, in such cases, be made a separate 'Section of the Works' with its own prescribed rate of damages for delay. If the main contractor is also late with another section of the works, there is no difficulty in separating the amount he should recover from his sub-contractor.

**8.2.4** Once aware of the problems, the situation in any particular contract may be helped in most cases by some modification or relaxation of the terms of payment the Employer agrees with his main contractor. He may agree to isolate the payments to nominated sub-contractors from the normal monthly payments for measured completed construction, or may agree to accept suppliers past results in lieu of actual performance tests, he may enter collateral agreements on certain obligations direct with the nominated sub-contractor or he may take action under Clause 59(A) of the ICE General Conditions, if they should be the ones applicable to the main contract. It may, however, never be possible to get the position completely clean, and the main contractor may have to appreciate what situation he might have found himself in if he had to let a normal sub-contract himself, and not been handed a nominated one.

**8.2.5** Nevertheless, it will by now be recognized that there are contractual dangers in having nominated sub-contracts, especially if the conditions of contract are not appropriate to both. Numerous legal cases have only served to highlight new dangers, rather than smooth the relationship. The troubles largely arise from the Employer having no direct contractual relationship with the sub-contractor, so that he can neither control his operations, nor sue him if he becomes in breach of his sub-contract. Any attempt to bridge the gap lessens the strength of the principle that the main contractor is wholly responsible to the Employer for the execution of the complete works.

Having said this, there are two more problems among the many which we must look at because their solution is not easy or straightforward.

*(a) Delay under ICE General Conditions (5th edition)*

When a delay situation occurs, due to the nominated sub-contractor being behind his contract completion date the Employer appears to be due payment of liquidated damages. However, under Clause 59A(6) of the ICE General Conditions (5th edition) he can only get from the main contractor such damages as he in turn can obtain from the sub-contractor for being late. But if the latter is in breach of his sub-contract, he is only liable for such damages as the other party (the main contractor) actually suffers, and as he does not have to pay his Employer anything, the amount is nil on both contracts.

Thus the Employer receives no payment for damages he undoubtedly sustains, and the cause of the trouble, the sub-contractor, gets away scot-free.

### (b) Design responsibility

In negotiating with a specialist firm with a view to a nominated sub-contract, the Employer settles with them the functional specification of the article he wants. His main contract is based on ICE General Conditions (5th edition), and under Clause 58(3) the main contractor takes no responsibility for design of the article nor the Employers agreed user specification. There is, therefore, no contractual link between Employer and nominated sub-contractor on these important aspects. The main contractor duly places his sub-contract. After delivery, it is found the article does not work properly even though it may meet the user's specification – the design is faulty. The Employer says: 'It's useless to me as it is; it must be changed.' The main contractor says: 'It's no business of mine. I have no design responsibility, and I ordered exactly to specification. Clause 58(3) is quite clear; I'm certainly not agreeing to it being reworded now, when everything has gone wrong.' The nominated sub-contractor says: 'It's true I discussed these matters with the Employer, but he brought up the idea, and certainly approved it. In any case, I have no contract with him – legally he doesn't exist as far as I am concerned. I have a valid order from the main contractor to supply an article to a user's specification, and this I've done. If its not what you want, too bad, but it's what you've got!'

This is the impasse that has to be sorted out, together with the allocation of cost for setting it right. A parallel case which, in practice, can easily lead to greater confusion, is if (after a main contract is in hand with the contractor being allotted no responsibilities under Clause 58(3)), the Engineer decides that a provisional sum included in the contract is to be used for a nominated sub-contract involving both design and specifications. The subject may be closely interwoven with the main contract works, but the main contractor is ignorant of the subject of the design and refuses to have Clause 58(3) altered. He claims, with justification, that such a change is not simply a variation the Engineer is entitled to make under Clause 51, as it can in no way be described as 'a variation to a part of the Works' but is a basic change of the relationships embodied in the contract.

These examples impress the need for the Employer, when negotiating with a firm in preparation for a nominated sub-contract, to insist on entering a collateral agreement direct with the firm as a pre-requisite to (and in consideration of) his nominating the firm to the main contractor. The agreement is framed to make the sub-contractor directly responsible to the Employer both for correctness of design and manufacture of the article, and for promptness in carrying out the contract to the programme agreed. The agreement might well be in the form of a model published by the Joint Contracts Tribunal (JCT) in support of their Standard Form of Building Contract.

## 8.3 RE-ISSUE OF ENQUIRIES AND RE-TENDERING

A word needs to be said about the undesirability of withholding for any reason the early award of a contract once tenders have been received and opened, and the subsequent issue of a fresh enquiry for almost the same job. Apart from the inevitable delay to the progress of the project, retendering is expensive (both to the tenderers and to the Employer), it can be most unfair to the original lowest tenderer and embarrassing, especially if, despite all security measures, the results of the previous competition have become known. It can also breed other undesirable consequences. It is preferable, whenever circumstances will permit, to negotiate modified requirements with the preferred tenderer (or tenderers) of the earlier enquiry – at least they have shown their superiority.

Retendering may, nevertheless, have to be considered if:

- The Employer feels obliged to make wholesale or fundamental changes in his requirements.
- The project has been seriously delayed or the contract programme substantially altered.
- There is evidence of serious collusion, favouritism or malpractice in connection with the first enquiry.
- No tenders received are up to the standard required, implying the need for a new specification or a new tender list (or both).
- It becomes impossible to accept a satisfactory tender before its validity (together with any extension the tenderer may be prepared to offer without alteration in price) expires.

- There has been a marked change in the economic picture, to the benefit of the Employer, since the original enquiry.

## 8.4   TENDER ACCEPTANCE

This subject has already been referred to in Section 1.2. The formal acceptance of the chosen tender and the setting up of a contract must be done by the Employer. The Engineer's role is one of advice to ensure that a proper and unambiguous relationship is duly established. This may include the preparation of any necessary documents in the Employer's name. In the simplest case where the Employer is accepting the offer of the tenderer – no more and no less – then a straight letter of acceptance completes the contract (Appendices 26A and B).

More usually, negotiations will have amended the original tender and the letter awarding the contract will only accept it subject to various changes since agreed. The letter of acceptance then constitutes a counter offer and itself needs a formal acceptance by the tenderer before a legally binding contract is concluded. To avoid any doubt the 'letter of acceptance' must call for such a reply. The date of the contract is determined by the date of whichever letter in the circumstances, *provides unconditional acceptance* of an offer that has been made by the other party (see Appendix 26C). Alternatively, if the negotiated changes to the tender are extensive it may be simpler to have the selected tenderer submit a revised tender embodying all the matters since negotiated, in which case a letter of acceptance from the Employer will then conclude the contract. In any case, it is advisable to record in writing at the time, the date when the contract actually came into force. If the Employer nevertheless requires the Engineer, as his agent, to issue a letter of acceptance, certain precautions have to be observed. These are referred to in Section 2.1.2 above, but briefly, the Engineer must ensure he has written authority from the Employer:

- To act as agent on his behalf.
- Stating precisely what he is empowered to accept (to accord with the Employer's intensions and to avoid repudiation or claims by the Employer later).
- Making it clear that the Engineer himself undertakes none of the Employer's obligations under the contract. The execution of deeds on behalf of the Employer is especially hazardous, and should always be avoided.

If the contract conditions prescribe that the contract is to be embodied in a formal Agreement under hand or under seal (the latter will normally be the case in large or complex contracts especially with public authorities and with overseas Employers) then a letter of acceptance is usually written to serve as a binding instrument until such time as the form of Agreement can be drawn up and signed or sealed, thus avoiding any last minute withdrawal by the contractor or any loss of time in the initial contract work. The contract dates from the date of the letter of acceptance, provided of course that it is unconditional.

## 8.5 INSTRUCTION TO PROCEED AND LETTER OF INTENT

**8.5.1** If for any reason it is required to establish a legal contractual relationship between Employer and contractor whilst still restricting temporarily the scope of activity to something less than the whole contract, this is done by issuing an *Instruction to Proceed* in place of the normal letter of acceptance. Typical reasons might be:

- The exclusion of certain matters still under negotiation or designs yet to be finalized.
- A temporary limit on the maximum financial commitment.
- A holding contract between Employer and a nominated sub-contractor until the formal sub-contract can be signed with the main contractor (work can then be started on the sub-contract in advance of the appointment of the main contractor).
- A holding contract pending incorporation into a formal Agreement.

An instruction to proceed must:

- Accept the tender in whole, in a specified part, or with stated exclusions.
- Specify a limit to which the contractor may go meanwhile (this can be financial, a part of the work etc.).
- Set out how (and if possible, when) the restriction is expected to be removed.

An instruction to proceed can be combined with a letter of intent for example: 'We intend placing the whole of this contract with you as soon as (specified negotiations) are completed but meanwhile will you please accept this instruction to proceed with (specified portions of the contract).' There are inherent difficulties in accepting only part of a tender, for example, what happens if the remainder falls out of validity before it, too, is accepted, or if the contractor changes his mind about accepting the rest of the contract. Exclusions are simpler to handle: '. . . accept the whole offer, except that Clause X shall not apply pending the outcome of current negotiations'.

By its nature, an instruction to proceed is rarely an unconditional acceptance of a tender. It therefore becomes a counter-offer requiring unconditional acceptance by the contractor in order to establish a valid contract. Once accepted it enables the contractor to proceed with the contract works subject to any limitations imposed. A typical example is shown at Appendix 25. Subsequently a further instruction or letter of acceptance is sent removing the limitations imposed by the instruction to proceed, such letter also requiring acceptance by the contractor. As with letters of acceptance, an instruction to proceed should preferably specify the salient features of the restricted contract it seeks to set up, for example:

- Contract documents.
- Limits of the restrictions imposed.
- Contract date.
- Contract price.
- Contractor's acceptance required.

**8.5.2** A *Letter of Intent* is of no legal value in the UK in establishing a contract and has little to commend it other than when some delay is foreseen in finalizing the contract documents. It gives the recipient due warning to make his plans for starting on the contract work smoothly and quickly, but any expenditure he may incur on the basis of the letter of intent is entirely at his own risk.

By 'letter of intent' we mean just what the name suggests: a letter which says in effect 'we intend placing a contract with you . . .'. It is, itself, clearly no acceptance. We mention this because there has grown a tendency in certain quarters to call a 'letter of intent' a document which, by virtue of its wording, is really an instruction to proceed. The distinction is important, as it might well affect the contract date (i.e. the date on which a valid contract was established) and matters dependent on it such as the onset of delay in completion, and liquidated damages.

The absence of legal value in the UK may not be true overseas. In some countries what amounts to a letter of intent is regarded as a binding undertaking such that the contract must subsequently be concluded as intended. Even in the UK a letter of intent has some moral value when issued by a firm of repute. Such a firm asking a contractor to start work on the strength of a letter of intent could only refuse to pay for any work done at peril of its reputation. A similar situation can arise if a contractor has, by agreement, to carry out design or development work to further negotiations with the Employer when setting up a negotiated contract.

## 8.6   TENDER REJECTION

Once details have been finalized with the preferred tenderers, any others may be rejected, but not before. No preferred tenderers on the 'short-list' should be rejected until a firm and binding contract has been established with the selected tenderer, in case of a last-minute hitch or withdrawal. Until a contract has been established the other preferred tenderers should not be told of the likelihood of success or rejection of their tenders, and indeed negotiations on outstanding points could, even at that stage, still be proceeding with at least one of them if, for example, the selected candidate is vacillating on a vital matter and his decision could make or mar his selection.

A typical letter of rejection is given at Appendix 27. It may be sent to the unsuccessful candidates by the Engineer once he has the approval of the Employer. It is usually handled by the contracts engineer.

One copy of each unsuccessful tender is usually retained by the Engineer for at least 6 months after the contract has been let, after which all copies can be destroyed.

# 9 *Matters outstanding after a contract has been signed*

## 9.1 MATTERS AFFECTING THE START OF WORK ON THE CONTRACT

**9.1.1** The start of work on a contract may depend on the successful conclusion of certain arrangements, often involving third parties which, by their nature, cannot be finalized until a firm contract has been established between the Employer and the contractor. The carrying out of the contract could nevertheless be prevented if the intended arrangements proved abortive. Such matters might include:

- Application for, and grant of, necessary import or export licences.
- Application for, and authorization of, payments in foreign currencies: (a) of the contract price by the Employers; (b) for sub-contracts by the main contractor.
- Approval in the UK of an overseas contract by the Export Credits Guarantee Department of the Department of Trade and Industry (ECGD) and the granting of insurance cover.
- Approval of terms of the contract by the foreign government of the Employer.
- Conclusion by either party of essential collateral agreements (e.g. consortium agreements, financial credit agreements, local authorities for the provision of public services to the contractor, agreements on rights of way or wayleaves).
- Provision of access roads, bridges, etc.
- Provision to the Employer/Engineer of undertakings, bonds or guarantees called for in the contract, and their approval of the terms and sureties offered.
- Payment of any sums due on, or immediately following, the signing of the contract.
- Establishment of any letters of credit or other financial instruments called for by the contract.
- Production of insurance policies required by the conditions of

contract, and approval of their terms and indemnities by the Employer/Engineer.
- Authority to proceed by the local planning authority.

The actions to be taken by the parties to the contract to negotiate and conclude such arrangements must be stated in the conditions of contract as obligations: failure to act once the contract is signed then constitutes a breach of a condition with all that that implies.

**9.1.2** Where such outstanding items exist, the contract must define an *effective contract date* in terms of the completion of outstanding items (see Section 4.10.2). If some of the outstanding matters fail to be settled satisfactorily, in spite of the best efforts of the parties, the contract itself may become void through impossibility of performance (e.g. no licences, no authority for foreign payments, etc.). The contract should specify these and lay down a minimum period which must elapse before the failure to settle them can be considered final, and the contract be voided. Otherwise an unhappy party might use a short half-hearted effort as a perfect excuse for a precipitate retreat from a contract.

If the Employer decides that the urgency of starting on the project programme cannot brook the delay up to the effective contract date, and ultimate settlement seems reasonably assured, he can issue an instruction to proceed to the contractor covering such preparatory matters as do not pre-empt a favourable decision by the third party. He accepts the risk that if the contract is voided he must pay for any work done within the limits of his instruction. If the contractor elects to start work without an instruction from the Employer, he has no right of recovery of his costs, if the contract has to be voided as a result of failure to agree the outstanding points.

**9.1.3** Since, by definition, the outstanding matters, such as those listed, can only be finalized after a firm contract has been agreed, they do not fall within the strict terms of reference of this book, namely the establishment of the contract itself. However, there are three items which are basically contractual in outlook and, for a successful conclusion, rely more on action by the contractor than by the third party concerned. They can be looked upon as cleaning up the negotiations which led up to the contract itself, and warrant further description:

- A formal Agreement confirming the contract.
- Provision of bonds and guarantees specified in the contract.
- Establishment of suitable insurance cover and indemnification of the Employer as specified in the contract.

## 9.2 FORM OF AGREEMENT

An intention by the Employer to confirm the contract by a formal Agreement must be included in the conditions of contract (preferably at the time of the enquiry) so that there is a firm obligation on the successful tenderer to do so. At the very least, a precautionary obligation 'if the Employer should so require' ought to be included, leaving the option with the Employer until he can see how much negotiation and change to his enquiry has to be tidied up and brought into the contract.

Confirmation of a contract in the form of an Agreement has four principal objects:

(a) To record formally that the contract is made by a corporate body, especially when all the tendering and negotiation has been made on its behalf by a single official or by its engineer. Indeed many public bodies are required by their standing orders to ratify, as being accepted and adopted by their elected council, a contract which has been arranged and signed by a permanent employee on their behalf. The recognized way of doing this is for them to conclude a formal Agreement, with the Council named as the contracting party.

(b) To tidy up a contract when negotiations have extensively modified the original enquiry/tender. The contract documents often become so involved that it needs a new schedule to be recorded, to show exactly what has been finally agreed.

(c) To emphasize the importance and formality of the contract.

(d) To extend the period during which a legal action or arbitration proceedings for any breach of the contract can be started, from the normal 6 years under the Statute of Limitations, to 12 years. For this to be achieved, the contract has to be in the form of an Agreement under seal (see below). Such an extension can be of great importance, for example in building contracts where defects in the contractors work may not become apparent within the 6-year period.

The form of Agreement is usually drawn up by the Employer with the advice and comment of the Engineer (or even to the recommendations of the Engineer, where his organization includes specialists able to draft it). Besides listing the contract documents, the Agreement often repeats the main features of the contract, the scope of the work, the contract price, the terms of payment, performance guarantees and so on. This detail is not necessary if all such points are clear in the contract itself. Many Agreements signed by local authorities purely to formalize the matter under their standing orders say very little other than (in suitable legal phraseology): 'This contract is approved by the Council.' Some standard forms of conditions of contract append a simple Agreement of this type, but they will frequently be found inadequate to meet the prime reasons for having an Agreement.

Agreements drawn up to ratify valid contracts already signed must be carefully drafted so that they conform exactly. Any new additions or exclusions will entitle either party to refuse to sign them, as being a new and different contract, but if they both do sign, then the terms of the Agreement will supplant those of the earlier contract from which they differ. Agreements are executed either 'under hand' (i.e. by the signatures of the two parties and their witnesses) or 'under seal', in which case the seals of the two parties are affixed and duly authorized by the signature of officials empowered by the party's constitution to do so. Where one or both of the parties are corporate bodies, execution will usually be under their corporate seals. Agreements under seal require to be officially registered at one of the eleven Inland Revenue Stamp Duty Offices in the UK, and under the Stamp Duty Acts they will emboss or affix a stamp recording the duty paid. For most engineering contracts the value is 50p. Registration has to be done within 30 days of the date of sealing by the second party, and there are arrangements for it to be done by post. Agreements under hand are less formal and differ in some legal effects: they require no registration or stamp.

Agreements are usually executed in two originals which are kept by the parties thereto. Stamping is only necessary on one of them, which is usually regarded as the Employer's copy. The Engineer should, if possible, arrange to have a further copy run off for easy reference and to keep with the master contract record (see Chapter 10).

## 9.3   BONDS AND GUARANTEES

To safeguard his position should certain eventualities arise during the course of the contract, it is not unusual for the Employer to include clauses in the conditions requiring the contractor to provide certain bonds and guarantees within a short period after the signing of the contract. These bonds and guarantees are not strictly parts of the contract documents. They deal with separate issues arising from the contract but external to it. They are separate contracts.

### 9.3.1   Bonds

A Bond is an undertaking by a party (in this case, the contractor) to do something for the benefit of another (in this case, the Employer). It is expressed in a Deed under seal. Undertakings by the contractor directly concerned with the contract are contained in the contract documents themselves, so bonds usually deal with external subjects. We have, for example, already encountered the *Tender Bond* which automatically becomes void as soon as a contract is signed. Another quite common undertaking is a *Repayment Bond*, often called for by an Employer when he agrees in the contract to pay part of the contract price in advance of the value being earned by the contractor (for example, a payment on signing, or the release of retention money before the end of the maintenance period). In the bond the contractor undertakes to pay the sum back if, for any reason he fails to earn it within a stated period.

A typical subject for a contractor's bond might be an undertaking to hold available for a stated number of years a full supply of spare parts for a non-standard machine supplied under the main contract; or, in a contract overseas, to make available at the overseas site specified specialist assistance to the Employer in the event he should find it necessary to put goods supplied by the contractor into store for a prolonged period under adverse conditions.

**9.3.2**   *A Guarantee* is an undertaking by a third party (*the Guarantor* or *the Surety*) to be answerable to a first party (e.g. the Employer) in a manner and to an extent specified in the guarantee, for a failure by a second party (e.g. the contractor) to perform a stated duty he owes to the first party. There may be a number of obligations legally taken by the contractor on which the Employer may need assurance that he will not be a loser financially if the contractor fails to carry them out. Any failure causing bankruptcy, for example, could make recovery of damages from the contractor himself very uncertain: it would be prudent in cases of doubt for the Employer to have the assurance of a reliable guarantor. Note, however, that a guarantee only comes into effect if the person owing the duty, actually fails to carry it out: it is not an alternative the Employer can elect to have in lieu of performance. Common eventualities viewed in this way are:

- Due performance: the contractor failing to fulfil his contract obligations, in whole or in part, or in respect of a stated vital aspect, such as a completion date.
- Warranty: the contractor failing to fulfil his obligations in respect of defective plant, or for proper maintenance. (A guarantee might be called for if the Employer agreed not to withhold any retention money as his own security).
- Subsequent contract: failure by a contractor to fulfil an undertaking to enter a further contract (e.g. an extension contract, a maintenance contract, etc.).

- Period of use: failure by the contractor to ensure the availability of replacement parts for plant, which he has undertaken will be for a minimum stated period.
- Performance of plant: failure to attain a minimum level of performance or efficiency by plant supplied under the main contract.
- Repayment guarantee: a guarantee instead of a repayment bond as described above. This would cover the case where the contractor did not repay because he had no money to do so.
- Contractors payments: especially overseas – failure by the contractor to pay (as provided in the main contract) certain local dues, such as, for example, local taxes, site-road upkeep, social-security dues for site labour, insurance premiums, etc.

**9.3.3** In most guarantees, the guarantor undertakes to pay to the Employer stated sums of money as his damages arising from the failure of the contractor to perform (see Appendix 28). The guarantor will rarely undertake specific performance, i.e. to carry into effect the actual duty of the contractor, and there are very good legal reasons why he will not. A possible exception might be the case of a parent company standing as surety for a wholly owned subsidiary over which it has complete operational control, that is to say it can bolster its finances, restructure its management, order it to employ specialist sub-contractors and in other ways require it to fulfil the obligations it has undertaken if necessary at the parent company's expense (see Appendix 29).

Such a parent-company guarantee is frequently called for when its subsidiary (the contractor) has only a nominal share capital and would not be able on its own to pay any substantial awards of damages, or indeed would not be able adequately to finance the execution of the actual contract. A guarantee by the parent ensures the latter would not let its subsidiary 'sink' rather than pay the cost of a rescue operation. An alternative procedure to such a guarantee would be for the Employer to place the contract with the parent itself, and include in its terms permission (or an obligation) for the work to be sub-contracted to the specialist subsidiary.

**9.3.4** The responsibility for accepting a bond or a guarantee relating to a contract in this way, and for the adequacy of its terms and provisions, rests with the Employer. The Engineer is not a party to it: his duties are advice and recommendation, but only to the extent that his facilities have allowed him to undertake such legal appraisals for his client. He might reasonably be expected to carry out a first-stage vetting and draw attention to such factors as he considered should receive expert legal advice. The matters he would be responsible for might include:

- Ensure the main contract conditions place an obligation on the contractor to provide such bonds and guarantees as the Employer requires, stating their main features.
- Nominate the contractor as responsible for proposing sureties for the Employers approval. Sureties will normally be specified as being a bank or recognized insurance company, or in some instances, a parent company.
- Check that the indemnities proposed are adequate, and related to the damages the Employer might expect to sustain from the contractors default. Allow for inflation.
- Specify the period within which the bond or guarantee must be furnished, and the period for which it must remain valid.
- Check the nature of the claim the Employer must make to receive payment, that is, how he proves that the contractor has failed (and is not just delayed!). This might be a certificate from the Engineer. Guarantors will rarely allow themselves to enter into arguments: they require a clear-cut and firm proof stated in the guarantee. If provided they pay; if not, they don't.
- Check as far as is within his powers, that the terms of the documents are adequate to express the failure covered and the circumstances which constitute failure. A very important point is that such bonds and guarantees tend to be worded around the contract as signed and hence might become invalid if the contract is changed in any way, for example by assignment, by extensions to the contract periods or by contract variations to the works. There must be a clause included in the guarantee to keep it in force whatever agreed changes occur, with or without the consent of the guarantor (see example in Appendix 28).

## 9.4 INSURANCE

**9.4.1** The Employer is interested from several points of view that the contractor has properly and sufficiently insured himself in regard to the contract and claims which might possibly arise from its execution. For this reason clauses are usually included in the conditions of contract specifying that certain insurances shall be maintained in force by the contractor, and the Employer shall have the right both to approve the policies and to be formally associated with them as a principal. The Employer's interests include:

- That the Contractor has fulfilled the legal requirements of the Employer's Liability (Compulsory Insurance) Act in respect of all employees working on the Employer's site. (*note:* The 'Employer' under the Act is the employer of labour (i.e. in our terms, the 'contractor). It is not, as in our terms, the party for whom work is being done under the contract.)
- That if, during the execution of the contract, the works suffer loss or damage, there will be no hitch in the contractor making them good and continuing with the execution due to shortage of money.
- That if the contractor has to meet claims from the public for loss, injury or damage arising in the course of the contract, the contractor will not be crippled financially by any such awards, and unable to fulfil his contract obligations.
- That in circumstances in which the Employer might be jointly sued with the contractor by a third party, the contractor will indemnify him to the extent specified in the contract, or to which the Employer has no direct contributory negligence. (For example, the Employer, as occupier of the site has a legal duty to safeguard the public using the area, and in the event that a third party is injured through the action of the contractor on the site, the Employer may still be jointly sued.)

**9.4.2** The normal contractual risks in engineering contracts fall into the scope of three basic forms of policy recognized by the insurance world. They are considered by insurers as three separate risks, often underwritten by different underwriters. In general terms the three types of policy are each fairly standard, though they can in most cases be negotiated in detail. The clauses in a contract dealing with insurance should, therefore, always have regard to the three divisions of risk and be worded so as not to conflict with what the insurance world would consider appropriate to each.

### 9.4.3 *Employer's liability insurance

This is insurance designed to meet the legal obligation in the UK of a contractor under the Employer's Liability (Compulsory Insurance) Act 1969. The contractor must insure against bodily injury or disease sustained by any employees under a contract of service or apprenticeship with him when such injury or disease arises out of and is sustained in the course of his employment. Note that it does not embrace self-employed

* See note above regarding definition of 'Employer' under the Act.

persons, the employees of sub-contractors, nor sub-contracted labour or hired-in operators of hired plant.

It is, however, often possible to negotiate the inclusion of the two last-named exceptions. Any of the excluded persons working on the site are treated as third parties under the contractors public liability insurance.

Cover under the employer's liability policy is usually unlimited but by law must be at least £2 million in any one occurrence.

### 9.4.4  Contract works insurance – 'Contractors' All Risks'

This insurance indemnifies the contractor against his liability for loss of or damage to the contract works whilst on the site (and, in the UK, in transit to the site) for so long as he remains at risk under the contract terms. Cover can sometimes be extended to include completed parts of the works for which the Employer has made some payment under the contract whilst they are still awaiting despatch from the fabricator's factory.

The risks covered may be limited to fire and special perils but are more usually specified as having to include the so-called 'all risks'. This is something of a misnomer, however, as the 'standard' all-risks policy will *exclude* a number of perils, such as:

- Damage due to wear and tear or caused to plant and machinery by its own failure or breakdown.
- Deterioration under environmental conditions reasonably to be expected at or around the site.
- Making good defective materials or workmanship or damage due to faulty design of the damaged part.
- Damage to works taken over by the Employer or taken into use or occupancy by him.
- Damage from pressure waves from aircraft or missiles.
- Loss or damage arising from war, invasion, nuclear explosions, nuclear radiations.
- The consequences of civil war, insurrection, usurped power, riots and civil commotion.

Loss and damage arising from strikes and lockouts are often bracketed with those from civil commotion, and the risk should be given careful consideration, as also should loss or damage arising from vandalism, collapse or subsidence, and theft. These risks can often be included, at the price of a somewhat higher premium, by negotiation with the insurers.

The amount insured is usually a minimum of the replacement cost of the works, the premiums being calculated initially on the value in the contract as signed. On completion of the works the premium is adjusted to allow for any increase in value due to variations to specification during the course of the contract, and any CPA applied.

In times of steep inflation, the full replacement value in the event of total destruction some time after original construction, might be well in excess of the original contract value. The insured should confer with his insurers as to how this increase can be best covered. Contract works are usually deemed to include temporary works in connection with the contract, and normally are extended to include also (under stated conditions and premiums) loss of, or damage to, constructional plant and machinery whilst on the site.

Contract works insurance policies usually specify an *insured's retained liability*, in one or a range of sums depending on the circumstances of a claim. This means that the insured himself must stand the first stated amount of any claim accepted by the insurers.

### 9.4.5 Public liability insurance

Public liability insurance protects the insured in respect of his legal liability for bodily injury to and loss of, or damage to, property of third parties. Most such policies will be subject to a number of exclusions and limitations of which the following are examples:

- Injury to his own employees whilst they are covered by the Employers Liability insurance.
- Loss or damage to the contract works.
- Loss or damage to the insured's own property or that in his charge or under his control.
- Making good defective or faulty workmanship or design.
- Consequences of traffic accidents on public roads.
- Consequences of explosion of boilers, pressure vessels.
- The use of or loss, injury or damage arising from the use of ship, aircraft or hovercraft.
- Liability arising from professional advice given.
- Liability voluntarily assumed under agreement.
- Loss or damage other than physical loss or damage to material property (e.g. ancient lights, patent rights, rights of way, are not covered by public liability insurance).
- Damage other than accidental (e.g. deliberate demolition would be excluded as well as normal expected risks attending it).

Note that the Employer's employees and property, those of other contractors on the site, self-employed persons and visitors to site would all fall under the cover provided to the contractor by his Public Liability policy, subject to the limitations and exclusions given above. Many of the exclusions listed are perils properly covered by other recognized insurance policies, for example, contract works, road vehicles, boilers.

The Employer will commonly prescribe in an enquiry the minimum amount his contractor will be required to cover in his public liability insurance. This in no way restricts the contractor's liability or risk, and he must still make his own decision as to how far he should insure himself. The figure quoted merely expresses the amount which would satisfy the Employer as being sufficient to cover his own interests and possible risks against which the contractor is required to indemnify him by the terms of his contract.

Contractors normally carry a public liability insurance in respect of all their business: they must consider whether any new contract imposes such greater risks and perils that the terms of their existing insurance need renegotiation to meet them.

### 9.4.6 Insurance clauses in contract conditions

The conditions of contract should include clauses setting out the contractor's obligations under the contract in respect of the two classes of insurance dealt with in Sections 9.4.4 and 9.4.5, treating each separately. The following points may need to be covered in each case:

- Contractor to maintain insurance cover in terms approved by Employer and with insurers acceptable to him.
- Contractor to insure up to a minimum of the sums specified and for a period continuously up to a stated date or stage in the execution of the contract.
- Contractor to pay all premiums as and when due.
- Contractor to produce policies and current premium receipts for inspection whenever required by the Employer/Engineer.
- If the contractor fails to maintain the stated policies, the Employer may do so on his behalf and deduct the cost of the premiums from the contract price.
- In the case of the contract works, contractor to devote payments received under the insurance policy to making good the damage in respect of which they were paid.
- The Employer's interest as principal to be recorded with the insurers and an 'indemnity to principals' clause to be included. Alternatively, the insurance to be in the joint names of the contractor and Employer as co-insured. Note, however, that if this latter alternative is applied to

a public liability insurance, the Employer *reduces* the extent of the protection he receives, unless the policy is also amended to include a so-called 'joint-insureds cross-liability clause'.

The contractors' obligations in the UK in respect of the Employer's liability insurance (Section 9.4.3) are, of course, statutory and need no reinforcement by a contract condition.

It is the Engineer's responsibility to advise the Employer on the terms of the clauses to be incorporated in the conditions of contract, and when the contract has been signed, to ensure the contractor duly arranges the policies in the form he is required to. The approval of the Employer is obtained and the receipts for the current premiums examined, both at the time and later, as they become due for renewal. The Engineer should also advise the Employer at the outset of any contract, of any special risks he considers are not properly covered by the contractor's policies, and for which the Employer should consider taking out insurance cover himself. Items, for example like third-party damage arising from demolition work or the use of explosives, losses through vandalism, the risk of ground subsidence, may not be adequately covered though they may be real perils none the less.

### 9.4.7 Insurance in contracts overseas

Contracts to be executed overseas need special consideration as regards insurance cover. It is a specialized field and the advice of a good insurance broker with appropriate experience should generally be obtained, especially if much overseas site work is involved. It must suffice here to mention a few of the more important aspects:

(a) Normal home policies as mentioned above are still required, and the Engineer must check whether the terms of the contractors existing policies extend to operations overseas without endorsement. For example:

- Reinstatement of damaged contract works may be a much more costly affair at a site overseas. Is the limit of the contractor's all-risks policy adequate in the circumstances? Does it cover exceptional local hazards – lawlessness, floods, termites etc. in so far as such perils are insurable.
- Is the contractor insured for employers liability in respect of expatriate employees working at site, and of local labour. What

about labour imported from a third country? Is it compulsory? Is cover adequate?

- Is his third party cover properly tailored to the local conditions, unusual risks, at the site? Does the country concerned have an equivalent of the statute of limitations limiting the period during which the contractor should remain insured against awards of damages? Or is his risk 'open-ended' in time?
- If labour is 'imported' from outside countries what are the contractors liabilities whilst they are in transit? Whilst living in labour camps near the site? When off-duty and off-site? For illness resulting from dietary or hygienic deficiencies, or from endemic diseases? Travelling on leave?

(b) The goods despatched by the contractor to site will need proper insurance during transit, i.e. by land, sea, and air from factory until unloaded on site. Full account must be taken of the conditions for landing at the specified port of entry (e.g. lighterage, demurrage, import procedure, delays) as well as the available means of transloading and movement across country to the site itself.

(c) It will frequently be found that contractors taking on engineering works in a foreign country are required by their conditions of contract to place all insurances having effect in the foreign country with a national or government-sponsored organization operating locally. Not only may the terms of these local policies be based on local law, but many give the contractor inadequate protection. With large contracts, the financial state and insurance expertise of some of these organizations may be inadequate for the job. A reliable UK broker should be required to arrange with the local body to reinsure the whole, or a large part, of his risks in the UK or in a market satisfactory to the broker (most of the local organizations spread their risks by reinsuring as a matter of routine, anyway). Any short-fall in the extent of the cover provided is then made up by insurance in London (the so-called 'Difference-in-Conditions' policy, covering the difference between the indemnity provided by the locally placed insurance and that needed for adequate protection by the Engineer and Employer). From the ECGD point of view, there are advantages in placing both the reinsurance and the DIC insurance in the London market.

These arrangements naturally cost money, and a tenderer must allow in his prices for the extra premiums he will incur. In some countries there is also a local tax on insurance policies which, though relatively small, should not be forgotten.

### 9.4.8  Insurance policies are legal documents

Insurance policies are specialized legal documents requiring considerable experience and expertise to interpret and analyse them. The Engineer may not have on his staff such a specialist, and consequently be able to offer the Employer only such limited comments as his experience allows. In the UK the Employer can obtain further help from the insurers themselves who, when provided with the relevant contract conditions relating to risk and insurance, will usually say if their policy meets the requirements, or if not, will specify alternative terms which do.

For his own security a contractor is well advised to get his insurer's views in this way on most major contracts: the Employer may well need to do the same before approving policies presented to him by the contractor. If a completely independent view is required, the policies and the relevant conditions of contract must be submitted to a qualified insurance consultant or experienced broker for his advice.

## 9.5  NOTIFICATION OF A CONTRACT COMING INTO FORCE

As soon as a contract is signed and comes into force, a considerable number of people, who are to be concerned with its supervision, recording and control, need to be informed of the fact and the salient features of the contract as finally agreed. They fall into two main groups.

**9.5.1**  The upper levels of management and administration concerned with the execution of the contract will have frequent need to refer to its exact terms, conditions and specifications so that they may be correctly interpreted and fulfilled. They need at hand a full and up-to-date set of contract documents as eventually agreed between the parties.

These must be prepared and issued to such individuals on a recorded 'numbered copy' basis, so that all subsequent amendments and additions can be properly issued to all holders (say, at monthly intervals). For this reason, if no other, the scale of distribution must be kept to a practicable minimum. Usually the distribution to all parties (Employer, contractor and Engineer) is organized by the contracts engineer on the Engineer's staff.

**9.5.2**  A wider group of executive officials concerned with the supervision, progressing and control of the contract need to be made aware that it is in force and the main points affecting its execution. A form 'Notice of Placing a Contract' containing such data can be issued immediately to all concerned so that their respective functions can be set in motion without delay. Suitable pro formae are suggested in Appendices 31A and B. Circulation to the appropriate list (see Section 2.1.4) needs no further comment.

**9.5.3**  One complete set of the contract documents, as signed, forms an important item in the master contract record (Chapter 10).

# 10  *The Master Contract Record*

## 10.1  THE PURPOSE OF THE MASTER CONTRACT RECORD

During the course of a contract and subsequent to it, matters of doubt and difference inevitably arise between the Employer, the Engineer, the contractor, his sub-contractors and others involved in the execution of the works. Settlement of the problem usually requires reference back to drawings, documents or records made earlier in the contract in order to establish the true and precise facts behind the matter. Disagreements which cannot be amicably resolved may well be put to the Engineer in his role as independent manager, or may find their way to a court of arbitration or a court of law. Reliable documentary evidence then acquires a new importance, and its timely retrieval can be a vital element in the preparation and prosecution of the case.

In such matters the Engineer has his usual dual responsibility. He was retained by the Employer to manage his project for him and, when faced with problems, the Employer has every right to turn to the Engineer to establish the facts. It is human nature that, since he has paid for the service, he will expect the Engineer to produce what he needs far quicker and in much more detail, then he would have been capable of doing himself, had he run his own project!

The Engineer has also his second role as impartial mediator between the parties in dispute, be they the contracting parties themselves or any sub-contractors or others who are prepared to accept his rulings as a more economic alternative to the formal courts. The onus is on the Engineer thoroughly to ascertain the facts and to be fair (and be seen to be fair – from the evidence) in his findings. His information must be accurate, complete, and immediately available: few things can be more embarrassing in such circumstances than for the Engineer to have a reasoned decision 'shot to pieces' by one of the dissidents producing a document he himself had failed to find.

## 10.2  RECORDS

The foregoing points to the necessity for a 'library' of all documents of importance which originate during the negotiations leading up to, and during the course of, the contract. It must be instituted by the Engineer and from it the project manager or the contracts engineer must be sure of finding without delay, anything pertinent to the question which has arisen. This 'library', the *Master Contract Record* is organized for each contract forming part of the Employer's project. It may be kept entirely by the contracts engineer, or (as is more usual in larger projects) in two parts:

- All technical records by the project engineer.
- All contractual, commercial and financial matters by the contracts engineer.

Whichever system is chosen, it must be clearly specified at the outset of the project by the project manager, and a proper system set in motion to ensure that copies of all documents having any relevance to the contract are routed to the person or persons holding the record.

Section 2.1.4 dealt with the plans for internal communications, and such plans must make sure that the master contract record is catered for. The routine suggested in Section 2.1.4 is suitable for the purpose and will help to reduce the rather thankless task of providing and filing the documents.

## 10.3  CONTENTS

The master contract record for any contract must be started with a separate identity as soon as the contract is let. A most important early item will be a full copy of the contract documents as signed (see Section 9.5), to include any letters of intent, instructions to proceed, letters of award and/or acceptance sent in reference to the contract.

It is also most advisable to include from the enquiry and tender files any earlier correspondence, records of meetings, contracts engineers pro forma diary (Section 3.7.3) and the like which show matters discussed or negotiated with the contractor at the tender stage. They define the intentions of the parties at the time and can prove very useful as the true history of a period which all too often has to rely on the memories of those who took part, or those who had the information second-hand.

Following the placing of the contract, the following classes of documents should be included in the master contract record (but the list will, to some extent, depend on the conditions and circumstances of the contract in question):

- Engineer's orders to start work (with dates).
- Copy of any formal Agreement confirming the contract.
- Copies of bonds, guarantees produced by the contractor.
- Copies of insurance certificates or reports dealing with contractor's insurance policies.
- Contract programme and progress charts established.
- Minutes of meetings, formal or informal, with records of approval thereof by those present.
- Site instructions issued.
- Engineer's instructions issued.
- Variation orders issued.
- All correspondence affecting or modifying the contract documents, specifications, drawings etc.
- Confirmations for important telex messages.
- Engineer's decisions, arbitrations or rulings (preferably with record of the basis on which they were made).
- Progress reports; updating of contract programme.
- All certificates issued in compliance with contract conditions.
- Claims from the contractor (including extensions of time, *force majeure*) and decisions made on them.
- Interim certificates passed for payment.
- Records of any delays to progress.
- Changes in officials appointed to the contract.
- Reports and supporting evidence on any accidents, losses, injuries, thefts, strikes or lockouts, vandalism, etc.
- Reports on and results of any inspections, tests, performance standards, etc., held or achieved.
- Record of any defects in the works and action taken to clear them.
- Record of contractor's activities during the maintenance period.
- Any contract documentation called for by any of the conditions of contract (e.g. shipping documents).
- Statement of all sums paid to the contractor by the Employer in respect of the contract price.
- Final statement of account and payment of balance.
- Record of any other matter occurring during the progress of the contract which might appear of value under the circumstances envisaged at the start of this chapter.

Much of the information stored in the master contract record may never be required, but when it is, it is required urgently.

*Appendices 1 to 31B*

## DEFINITIONS

To aid precision in interpreting a contract it is usual to begin the conditions of contract with a list of definitions of the terms used. Some common words are allotted a narrower, more special meaning than usual. When, in the documents, the word is intended to have this special meaning it is spelt with a capital initial letter: when intended to have its normal dictionary meaning it is used with a small initial letter.

The following is a list of typical examples; there is, however, no standardization, so that when studying a contract document, reference must be made to the list contained in it. Corresponding terms used in some standard forms of conditions are compared in Appendix 2.

*The Employer:* The person ordering the goods, works or services covered by the contract, including his legal representatives, assigns or successors.

*The Contractor:* The person whose offer to provide goods, works or services has been accepted by the Employer, including the contractor's legal representatives, assigns or successors.

*The Engineer:* Shall mean ........*(name)*........ or the person for the time being or from time to time notified in writing by the Employer to the Contractor as Engineer for the Contract, or in default of any notification, the Employer.

*The Contract:* Shall mean the bargain agreed between the Employer and the Contractor for the execution of the Works including all documents to which reference may properly be made in order to ascertain the rights and obligations of the parties to the said bargain.

*Contract Price:* The sum to be ascertained and paid in accordance with the terms of the Contract for the execution of the Works by the Contractor.

*Contract Value:* The valuation on the basis of the Contract Price of a stated portion of the Works in the condition and at the place in which that portion is at the relevant time.

*The Works:* All Plant to be provided and works and services to be done by the Contractor under the Contract.

*Plant:* All machinery, apparatus, articles, materials and things to be provided by the Contractor under the Contract other than Contractor's Equipment.

*Permanent Works:* All work to be permanently constructed and completed for the Employer under the Contract.

*Temporary Works:* All temporary work of every kind required in or about the construction of the permanent works but not for incorporation therein.

*Contractor's Equipment:* All tools, tackles, stores, machinery, vehicles, apparatus, articles, materials and things brought to site by the Contractor for the purpose of executing the Contract but not for incorporation in the Works.

*The Site:* The place where Plant is to be delivered or works to be carried out by the Contractor under the contract together with so much of the surrounding area as the Contractor shall be allowed to use in connection with the Works other than purely for the purpose of access to the place.

*Section of the Works:* An identifiable portion of the Works so specified and described in the contract documents.

*Day, Month and Year:* Shall be the same according to the Gregorian calendar.

*FOB, C and F, CIF, etc.:* Shall be as defined in INCOTERMS 1953 (revised 1977).

*Writing:* Shall include any manuscript typewritten or printed statement under hand or seal as the case may be. In some contracts, a telex from an authorized signatory is accepted as 'writing' but this should be checked. Not cables.

The terms shown in any line have generally the same meaning but there are cases where the definitions and use are not identical. The terms for FIDIC (Civil) and FIDIC (Electrical and Mechanical) Conditions are the same as those used in the ICE (5th edition) and IMechE/IEE Model Form 'A' Conditions, respectively.

| ICE (5th Edition) | IMechE/IEE 'A' | IChemE | Standard Form of Building ('RIBA' or 'JCT') | EEC (574A) | BEAMA 'B' | This volume |
|---|---|---|---|---|---|---|
| Employer | Purchaser | Purchaser | Employer | Purchaser | We | Employer |
| Contractor | Contractor | Contractor | Contractor | Contractor | You | Contractor |
| Engineer | Engineer | Engineer | Architect/Supervising Officer | – | – | Engineer |
| (Temporary) Works (Permanent) Works | Works | Works | Works | Works | – | Works |
| – | Plant | Plant | 'materials and goods' | Plant | Goods | Plant |
| Constructional Plant | Contractor's Equipment | Contractor's Equipment | – | Contractor's Equipment | – | Contractor's Equipment |
| Completion | Completion | Completion | Practical Completion | Completion | Completion | Completion |
| – | Taking Over | Taking Over | 'taking possession of' | Taking Over | Taking Over | Taking Over |
| 'final test' | Tests on Completion | Take-Over Tests | – | Take-Over Tests | – | Tests on Completion |
| Certificate of Completion | Taking Over Certificate | (Taking Over Certificate) Acceptance Certificate | Certificate of Practical Completion | Taking Over Certificate | – | – |
| 'a certificate' | Interim Certificate | 'a certificate' | Interim Certificate | – | – | Interim Certificate |
| Maintenance Certificate | Final Certificate | Final Certificate | Final Certificate | – | – | Final Certificate |
| Period of Maintenance | (no specific term) | Defects Liability Period | Defects Liability Period | Guarantee Period | – | Period of Maintenance |
| Contract Price | Contract Price | Contract Price | Contract Sum | – | Contract Price | Contract Price |
| – | Contract Value | – | – | – | Contract Value | Contract Value |

Consulting Engineers

# *Truebill, Klinch & Steer*

Ref ..................

Project Management Division
2001 Victoria Street
London SW1   England

Telephone (01) 333–3333
Cables Clincher Westend
Telex 1234567 (Reply back  −  Trustern)

To ........................................................

..........................................................

Date ................

Dear Sirs,

We would be obliged if you could provide us with data on the lines indicated about your company and its activities, for inclusion in our internal register of suppliers, producers and contractors in the engineering field. The register is solely for our own use in our engineering consultancy operations.

For and on behalf of Truebill, Klinch & Steer

..........................................................
Contracts Engineer

| *Suppliers, Producers and Contractors in the engineering field:* *Form of enquiry for information* | |
|---|---|
| 1. (a)  Full name of company | (a) |
| (b)  Registered address | (b) |
| (c)  Telephone and STD code | (c) |
| (d)  Telex No. and Reply | (d) |

| | |
|---|---|
| 2. (a) *Type of company* <br><br> (b) Status <br><br> (c) Capital <br> (d) Debentures/Loans <br><br> (e) Bankers | (a) Private/Exempt Private/Public/Limited Liability/Partnership <br> (b) Independent/Fully-owned subsidiary/ Associate/Parent <br> (c) Authorized £.............. Issued £.............. <br> (d) £.................... maturing ..................... <br> Interest rate ............% p.a. <br> (e) Name: .................. Branch: .................. |
| 3. *Associated Companies* (state parent, associate, technical liaison, etc., and their activities) <br><br> (use separate sheet if necessary) | (a) <br><br><br> (b) <br><br><br><br> (c) |
| 4. *Type and range of work you are able to undertake* (send leaflets, brochures, etc., as available) | |
| 5. *Turnover* (last 3 years) on the work described in 4 above. <br> If available include copies of recent balance sheets (Profit/Loss and Assets/Liabilities) and any other financial data you think might be useful | Year     Turnover     Profit (after tax) <br> (a) 19–/– £ .................... £ .................... <br><br><br> (b) 19–/– £ .................... £ .................... <br><br><br> (c) 19–/– £ .................... £ .................... |
| 6. *Geographic areas of operation* <br>     (a) In UK (state branches) <br>     (b) Abroad (state countries) <br>     (c) How are you represented abroad? | (a) <br><br> (b) <br><br> (c) |

| | |
|---|---|
| 7. *Employees* (numbers)<br>Average last 3 years of:<br>*Staff*<br><br><br><br><br><br><br><br>*Labour Qualifications* | (a) Contract, technical and management ............<br>(b) Contract administration ............<br>(c) Design and drawing office ............<br>(d) Research and development ............<br>(e) Company management and administration ............<br>(f) Average full-time labour force employed ............<br>(g) Percentage of (a) + (b) who are engineering graduates ............%<br>(h) Percentage of (a) + (b) who are members of recognized engineering institutions ............% |
| 8. *Constructional Plant*<br>List by major categories of constructional plant owned by the company and available for project work | |
| 9. *Contracts recently completed*<br>(a) Main project title:<br>............................<br>............................<br>Employer:<br>............................<br><br>(b) Main project title:<br>............................<br>............................<br>Employer:<br>............................<br><br>(c) Main project title:<br>............................<br>............................<br>Employer:<br>............................ | Contract Work ....................................<br>Period ......../........ Value £ ....................<br>Location ..........................................<br>Contract Engineers/Consultants ....................<br>..................................................<br><br>Contract Work ....................................<br>Period ......../........ Value £ ....................<br>Location ..........................................<br>Contract Engineers/Consultants ....................<br>..................................................<br><br>Contract Work ....................................<br>Period ......../........ Value £ ....................<br>Location ..........................................<br>Contract Engineers/Consultants ....................<br>.................................................. |

| 10. *Trade Associations* to which your company belongs | (a) |
| | (b) |
| | (c) |
| 11. *Any further data* you consider relevant. | |

Date:..................... Signed: ...............................................................
Position: ...............................................................

Consulting Engineers                                              **CONFIDENTIAL**

## *Truebill, Klinch & Steer*                         Ref .................

Project Management Division
2001 Victoria Street
London SW1   England

Telephone (01) 333–3333
Cables Clincher Westend
Telex 1234567 (Reply back — Trustern)


*(Addressee)*
.......................................................                 Date .................

.....................................................


Dear Sirs,

<u>Preliminary enquiry</u>

Tender for.......................................

On behalf of our client ............ we will shortly be inviting tenders for the supply/
construction/erection of ............ including ............ at ........*(location)*........ As an
indication of the size of the contract we can state that it will involve ............ (here
state approximate size, number, production rates or the like, but avoid any mention
of approximate cost).

The enquiry will be issued on or about ........*(date)*........ and we expect tenders to be
returnable by ........*(date)*........

It is anticipated that the date for delivery/start of work on site will be
........*(date)*........ and completion of construction/erection by ........*(date)*........
based on the assumption that a firm contract will be placed with the successful
tenderer by ........*(date)*........

Would you kindly inform us as soon as possible (and in any case not later than
........*(date)*........) if you wish to be considered for this work. An affirmative reply
must not, however, be taken as an assurance that you will be included in the tender
list. If you do not wish to submit a tender in this instance it will not preclude you from
being invited to do so in the future.

(Optional):

In order that our records may be brought fully up to date, we enclose herewith our standard Enquiry Form for Information and invite you to complete and return it to us whether or not you wish to submit a tender in the present instance: the information requested will be treated by us in full confidence.

Yours faithfully,
for and on behalf of
Truebill, Klinch & Steer


..............................................................
Project Manager

# *Truebill, Klinch & Steer*

Project Management Division        Internal Memo dated ...............................

## Tender Enquiry Programme

Employer ......................................        Enquiry no. ....................................
Project .........................................        For ..............................................
Issued by.......................................        Issue no. ......................................

*For action to*

Employer (a) ...............................
Employer (b) ...............................
Employer (c) ...............................
Project Engineer
Quantity Surveyor
(or estimating department)
Engineering Design (Civil)
Engineering Design (E & M)
Contracts Engineer

*For information to*

Employer (d) ................................
Employer (e) ................................
Employer (f) .................................
Project Manager
Project Planning Engineer
Engineer's Representative

The timetable for the above project requires that the enquiry for the contract/sub-contract shown in the heading must be issued shortly. The following programme must be adhered to in its preparation. It is important that all concerned meet the key dates shown.

PROGRAMME

| Action | Person responsible | Date for completion |
|---|---|---|
| 1. *Tender list* <br>   (a) Suggested names for inclusion to contracts engineer <br>   (b) Preliminary list approved <br>   (c) Preliminary enquiries issued <br>   (d) Preliminary replies received <br>   (e) Status investigations completed <br>   (f) Final tender list assembled and approved by Engineer <br>   (g) Final tender list approved by Employer | | |

| Action | Person responsible | Date for completion |
|---|---|---|
| 2. *Specification and broad description of the works*<br>(a) Draft completed<br>(b) Employer's approval of draft obtained<br>(c) Final draft completed, typed and checked<br>(d) Final draft copies to quantity surveyor and contracts engineer<br>(e) Enquiry drawings complete<br>(f) Employer's approval of drawings obtained<br>(g) Copies as required to quantity surveyor for taking off<br>(h) Velos to contracts engineer for printing | | |
| 3. *Bills of quantities*<br>(a) Quantities taken off by quantity surveyor<br>(b) Draft bills typed and checked<br>(c) Preamble to bills complete and agreed with contracts engineer<br>(d) Provisional and prime cost sums estimated and agreed<br>(e) Contract value estimated and compared with control estimate<br>(f) Masters of preamble and bills to contracts engineer for reproduction | | |
| 4. *Form of Tender*<br>(a) Commercial and financial proposals approved by Employer (including rates of liquidated damages)<br>(b) Contract conditions and site regulations compiled<br>(c) Contract programme finalized<br>(d) Instructions to tenderers and draft form of tender approved by Employer<br>(e) Master copies available to contracts engineer for reproduction | | |
| 5. *Enquiry documents*<br>(a) All master documents updated ready for printing by ....................:.<br>(b) Printing and collation complete (including drawings) | | |

| Action | Person responsible | Date for completion |
|---|---|---|
| (c) Document covers designed<br>(d) Document covers printed and available<br>(e) Document sets assembled and packed ready for issue<br>(f) Enquiry documents issued to tenderers | | |
| 6. *Tender appraisal*<br>  (a) Tenders due for return by ............<br>  (b) Tender opening due<br>  (c) Tender and bills of quantities checked arithmetically<br>  (d) Appraisal reports available to contracts engineer<br>    by project engineer<br>    by finance<br>  (e) Draft appraisal report ready<br>  (f) Negotiations with preferred tenderers finalized<br>  (g) Appraisal report and recommendations complete and approved by project manager. Sent to Employer | | |
| 7. *Decision*<br>  (a) Selection of contractor made<br>  (b) Employer confirms sanction of capital expenditure accordingly<br>  (c) Letter of Acceptance/ITP sent to successful tenderer<br>  (d) Start date on site<br>  (e) Letters of rejection sent to unsuccessful tenderers | | |

## Contract enquiry – daily log

Project ..................................................... Date ................................

Enquiry for ............................................... Sheet serial No. ...................

Enquiry reference No....................................

*PART 1: Actions due this day*

| Actions due this day | Action by | File reference | Date action taken |
|---|---|---|---|
| | | | |
| | | | |
| | | | |
| | | | |
| | | | |

*PART 2: Record of unprogrammed events*

| Serial No. | Time | Event | Action taken by contracts engineer |
|---|---|---|---|
| | | | |
| | | | |
| | | | |
| | | | |
| | | | |

# *Truebill, Klinch & Steer*

Project Management Division          Date ................................................

## Record of enquiries issued

Issued by....................................................

| Employer ...................................... | Tender for ...................................... |
| Project ...................................... | Enquiry reference No: ...................... |

Documents issued to each tenderer comprise
(a)  ............ copies of Instructions to Tenderers
(b)  ............ complete sets of enquiry documents in ...... vols
(c)  ............ complete sets of ...... drawings
(d)  ............................................ copies of ......................................
(e)  ............................................ copies of ......................................

| No. | Name and address of tenderer | Date of despatch | Method of transmission |
|-----|------------------------------|------------------|------------------------|
| 1 | | | |
| 2 | | | |
| 3 | | | |
| 4 | | | |
| 5 | | | |
| 6 | | | |
| 7 | | | |
| 8 | | | |

Consulting Engineers

# *Truebill, Klinch & Steer*

Ref ...................

Project Management Division
2001 Victoria Street
London SW1   England

Telephone (01) 333–3333
Cables Clincher Westend
Telex 1234567 (Reply back  —  Trustern)

*(Addressee)*
.................................................................

Date .................

.................................................................

Dear Sirs,

*(Employer and Project)*
.........................................................................................................

Enquiry No. ..................................... for .....................................

We are instructed by our client, the Employer, to invite you to tender for the above work and for this purpose we enclose ............ sets of the enquiry documents and the Instructions to Tenderers (which lists the enquiry documents in its opening paragraph).

Your tender, which must be made and submitted strictly in accordance with the Instructions to Tenderers, is required to be in our offices at the above address by 2 p.m. on ............ Tenders received after this time (or any authorized extension thereof which we might notify later) will not normally be considered.

For the submission of your tender we enclose a supply of pre-addressed tender labels. These must be used on all packages you submit and no other marks or titles must appear thereon which in any way indicate that the package has been sent by you.

If you have any queries on the enquiry documents, these must be made in writing or by telex to our contracts engineer (Mr ............).

Please acknowledge, by return first-class mail, your receipt of this letter and all the enclosures listed, confirming that you expect to submit a tender by the appointed time. This receipt is required even if you have collected the enquiry documents from our office and your messenger signed a receipt for them at the time of collection. If you do *not* intend to make an offer please return all documentation to us without delay, together with your letter of refusal.

Yours faithfully,
for and on behalf of
Truebill, Klinch & Steer

.................................................................
Contracts Engineer

## Tender return label

---

# <u>TENDER</u>

***DELIVER UNOPENED TO:***

(1)     MR........................ Dept........................
          CONTRACTS ENGINEER (PM DIV)
          TRUEBILL, KLINCH & STEER,
          2001 VICTORIA STREET,
          LONDON SW1

          ENGLAND

---

(2)  **Enquiry No. ......................  For ......................**

(3)  **Return before** ........*(time)*........ **on** ........*(date)*........

---

Above is a suggested layout for a return label to be sent to all tenderers with an enquiry. The quantity sent should suffice for all tender documents to be returned in parcels of reasonable size.

Before despatch to tenderers, details on the label must be inserted by the Engineer as follows:

(1) Name and department reference of the contracts engineer to whom the parcel is to be delivered.
(2) Identity of the contract for which the tender has been sent (in case several enquiries are out to tender at the same time)
(3) Date and hour by which tenders have to be in the hands of the Engineer.

Labels must be used on all parcels or packages containing tender documents or drawings, even if they are being delivered by hand of messenger.

No other marks which might reveal the tenderers identity must appear on the outside of the packages or parcels.

......................................................
*(Insert Employer's)*                                    Enquiry ref. no. ............
......................................................
*(name and address)*
......................................................

......................................................

Dear Sirs,

## Declaration of bona fide competitive tender

*(Employer/Project)*
......................................................................................................................
*(Tender for)*
................................................................ Tender ref. ................

The essence of selective tendering is that the Employer shall receive bona fide competitive tenders from all firms tendering. In recognition of this principle, we declare that this is a bona fide tender, intended to be competitive, and that we have not fixed or adjusted the amount of the tender by or under or in accordance with any agreement or arrangement with any other person. We further declare that we have not done and we undertake that we will not do, at any time before the returnable date for this tender, any of the following acts:

(a) Communicating to a person, other than the person calling for these tenders, the amount or approximate amount of the proposed tender.

(b) Entering into any agreement or arrangement with any other person that he shall refrain from tendering or as to the amount of any tender to be submitted.

(c) Offering or paying or giving or agreeing to pay or give any sum of money or valuable consideration directly or indirectly to any person for doing or having done or causing or having caused to be done in relation to any other tender or proposed tender for the said work any act or thing of this sort described above.

In this declaration, the word 'person' includes any persons and any body or association, corporate or incorporate; and 'any agreement or arrangement' includes any such transaction, formal or informal, and whether legally binding or not.

Date ...........................................     Signed .........................................

Position .......................................

For and on behalf of .........................

......................................................

191

## THE FORM OF TENDER

1. This is a letter addressed to the Employer in which the tenderer formally makes his offer. It is usually issued as an enquiry document, for completion by the tenderer, in order to avoid the many differences and ambiguities which occur if each tenderer is allowed to compose his own. It fixes a uniform and satisfactory basis of offer for them all.

2. The first paragraph enumerates all the documents to which the offer is to be subject. Besides the tender price there are usually other figures and facts which each tenderer is required to provide: forms are normally provided for them, and they are conveniently introduced into the list of documents by naming each a 'Part' of the Form of Tender. This avoids a long recital of titles. Thus for example, a Form of Tender might comprise 'Parts 1–7' which, written in full might be:

   Part 1: The Letter of Offer.

   Part 2: Breakdown of tender prices.

   Part 3: Contract programme and scales of liquidated damages for delay in meeting it.

   Part 4: Schedule of rates (for pricing variations).

   Part 5: Schedule of sub-contractors proposed.

   Part 6: Recommended list of spare parts to be held.

   Part 7: Schedule of all items of non-compliance of tender with enquiry.

3. Signing the Letter of Offer (Part 1) automatically embraces as parts thereof, all the other documents so introduced, including the other Parts of the offer. They do not need to be all signed individually, though it is customary for the person signing the letter of offer to initial them all as a means of identification. Of course, any form not introduced into the letter of offer in this way but left to stand on its own needs to be signed by the same person signing the offer.

4. This system has been adopted in the examples of typical letters of offer given in this Appendix 10. Three such typical examples are shown, appropriate to three different types of contract.

   Appendix 10A: For a remeasurement contract with appropriate bills of quantities, where a single specific tender price would not itself be meaningful.

   Appendix 10B: For all tenders in which a firm tender price, lump sum, is a meaningful offer.

   Appendix 10C: for tenders intended to be incorporated as a nominated sub-contract with a main contractor who has not yet been appointed.

## APPENDIX 10A  *Form of Tender — Part 1: Contract subject to remeasurement*

To ................................................  Tender No. ...............................

   *(Insert Employer's)*
.............................................  Date ........................................

   *(name and address)*
.............................................

.............................................

Gentlemen,

   *(Project)*
......................................................................................................

   *(Enquiry title and ref. No.)*
......................................................................................................

1. Having examined the Form of Tender Parts 1, 2, 3, 4, 5 and 6 and the appendices thereto, the Conditions of Contract, the Special Conditions, the Site Regulations, the Specifications, Drawings and Bills of Quantities for the above-named contract works, we offer to construct, erect, complete and maintain the whole of the said works in conformity with the above-mentioned documents for such sum as may be properly ascertained in accordance therewith and the prices, rates, terms and conditions we have inserted therein.

2. The offer contained in this tender will remain valid and open to acceptance by you for a period of ............ days after the due date for return of tenders specified in your enquiry (or such later date for return as may have been subsequently notified by you).

3. We hereby declare we have priced our tender at the prices ruling at ........*(date)*........, and in accordance with any legislation in force on that date.

4. If our tender is accepted we will, when requested, provide a Bond of a bank or insurance company in terms to be approved by you (which approval shall not be unreasonably withheld) to be jointly and severally bound with us in the sum specified in the aforesaid conditions of contract for the due performance by us of all our obligations and liabilities under the contract.

5. Our formal declaration in the terms required by you that this is a bona fide competitive tender accompanies this Form of Tender — Part 1.

6. Unless and until a formal Agreement has been executed by us both, this tender with your written acceptance thereof shall constitute a binding contract between us. We understand that you are not bound to accept the lowest or any tender.

Signed .............................................................................

In the capacity of ................................................. duly authorized to sign

tenders for and on behalf of .................................................................

.................................................................

193

To ........................................................  Tender No. ................................

*(Insert Employer's)*
........................................................  Date ........................................

*(name and address)*
.................................................................

.................................................................

Gentlemen,

*(Project)*
.................................................................................................

*(Enquiry title and ref. No.)*
.................................................................................................

1. Having examined the Form of Tender Parts 1, 2, 3 and 4 and the appendices thereto, the Conditions of Contract, Special Conditions, Site Regulations, Specifications and Drawings for the above-named contract works issued by you with your enquiry we offer to supply, deliver, erect, commission, complete and maintain the whole of the said works in conformity with the above-mentioned documents further defined and illustrated by our technical descriptions and drawings annexed to this tender and described herein as forming part of our offer, all for the sum of ............ Pounds Sterling (£Stg............) or such other sum as may be properly ascertained in accordance with the terms and conditions of the contract.

2. The offer contained in this tender will remain valid and open to acceptance by you for a period of ............ days after the due date for return of tenders specified in your enquiry (or such later date for return as may have been subsequently notified by you).

3. We hereby declare that we have priced our tender at the prices ruling at ........*(date)*........, and in accordance with any legislation in force on that date.

4. We understand that you may at your discretion place orders for any portions of the works separately itemized and priced in Part ............ of the Form of Tender with different tenderers and we confirm that except to the extent stated in the said Part ............ the prices we have quoted for any portions ordered from us will not be altered by such action.

5. If our tender is accepted we will when requested provide a Bond of a bank or insurance company in terms to be approved by you (which approval shall not be unreasonably withheld) to be jointly and severally bound with us in the sum specified in the aforesaid conditions of contract for the due performance by us of all our obligations and liabilities under the contract.

6. Our formal declaration in the terms required by you that this is a bona fide competitive tender accompanies the Form of Tender – Part 1.

7. Unless and until a formal Agreement has been executed by us both this tender with your written acceptance thereof shall constitute a binding contract between us. We understand that you are not bound to accept the lowest or any tender.

Signed .........................................................................

In the capacity of ................................................... duly authorized to sign

tenders for and on behalf of .................................................................

.............................................................................

To .................................................

*(Insert Employer's)*

.................................................

*(name and address)*

.................................................

.................................................

Tender No. ...............................

Date .......................................

Gentlemen,

*(Project)*

.............................................................................................................

*(Enquiry title and ref. No.)*

.............................................................................................................

1. Having examined the Form of Tender Parts 1, 2, 3 and 4 and the appendices thereto, the references to and extracts from your proposed contract for ............ ('The Main Contract') the Specification and Drawings describing the sub-contract works all as included in your sub-contract enquiry we offer to supply, deliver, erect, commission, complete and maintain all the said sub-contract works in conformity with the above-mentioned documents further defined and illustrated by our technical descriptions and drawings annexed to this tender and described herein as forming part of our offer all for the sum of ............ Pounds Sterling (£Stg............) or such other sum as may be properly ascertained in accordance with the terms and conditions of the sub-contract.

2. We undertake that if required by you we will enter into negotiations with a contractor you shall name as your Main Contractor for a nominated sub-contract for the purpose of executing and completing the sub-contract works aforesaid for the sums hereinbefore mentioned. As and when such nominated sub-contract shall have been concluded with the said Main Contractor, any contract that may exist between us as a result of your acceptance of the present tender shall be deemed to have been replaced by the said nominated sub-contract and shall become null and void. Such expenditure (if any) as we shall have incurred with your authority in relation thereto shall at your decision either be paid to us and the amount deducted from the sub-contract price to the Main Contractor or shall be charged by us to the latter.

3. We are prepared to negotiate with you a collateral Agreement generally on the lines of the model published by the Joint Contracts Tribunal in which we retain responsibility to you direct for our design commitments and for the prompt execution of the sub-contract works even though the provisions of the previous clause 2 shall have been put into effect.

4. The offer contained in this tender will remain valid and open to acceptance by you for a period of ............ days after the date of this tender as given above.

196

5. We hereby declare that we have priced our tender at the prices ruling at ........*(date)*........, and in accordance with any legislation in force on that date.

6. Our formal declaration in the terms required by you that this is a bona fide competitive tender accompanies the Form of Tender — Part 1.

7. Unless and until a formal Agreement shall have been executed by us setting out the terms of this offer and your acceptance of it, this tender itself together with your acceptance of it in writing shall constitute a binding contract between us. We understand that you are not bound to accept the lowest or any tender.

8. We undertake that if, during the negotiations with the Main Contractor as provided in Clause 2 above, we are so requested by him, we will provide a Bond of a bank or insurance company in terms to be approved by him (such approval not being unreasonably withheld) to be jointly and severally bound with us in the sum specified in the aforesaid extracts from the Main Contract for the due performance by us of all our obligations and liabilities to him under the sub-contract. The provisions of Clause 3 above shall not be affected by this undertaking.


Signed ...........................................................................

in the capacity of ................................................... duly authorized to sign

tenders for and on behalf of ...............................................................

...........................................................................

*(Tender for)*
............................................................................................................

## · FORM OF TENDER – PART ........

## Data relating to authorized overtime and daywork

Tenderers are required to complete fully the data required by the following forms and paragraphs and to submit them with their tender. If the information called for is not supplied or is incomplete, the whole tender may be invalidated.

### 1. *Normal working hours*

Tenderers shall complete Table 1. It shows the starting and finishing times each day between which the tenderer expects to work and *for which he has allowed and included in his tender price and rates*.

Overtime will not be authorized by the Engineer for reimbursement in respect of any period between the daily times shown.

TABLE 1

|  | Starting time a.m. | Finishing time p.m. |
|---|---|---|
| Monday | | |
| Tuesday | | |
| Wednesday | | |
| Thursday | | |
| Friday | | |
| Saturday | | |
| Sunday | | |
| No. of Saturdays per four week period<br>No. of Sundays per four week period | | |

### 2. *Basic labour rates*

Tenderers shall complete Table 2 showing for each category of labour it is proposed to employ on the works the basic rate payable to the man (on which the tender price is based). 'Basic rate' shall include craftsmen's plus rates *but shall exclude* all bonuses, allowances, travelling and subsistence expenses, weekly supplements, rates paid for conditions and the like.

198

The rates shown (in accordance with the appropriate national agreement) shall be those current at the contractual base date of the Price Adjustment Clause in the contract (if any) or otherwise at the date of tender.

TABLE 2

| Labour category | Basic rate (£/hr) | Working rule agreement used |
|---|---|---|
| (1) | (2) | |

3. *Overtime rates*

Tenderers shall complete Table 3 to show the periods (immediately before or following the times given in Table 1 above) during which different overtime increments are payable.

The hours and increments inserted must be in accordance with the Working Rule Agreement either of the Civil Engineering Construction Conciliation Board for Great Britain or of the National Joint Council for the Building Industry or of such other body recognized as appropriate for negotiating the wages and conditions of employment of labour in specific trades not covered by the agreements referred to above, whichever is properly applicable to the work.

TABLE 3

| | Hours — time and a half | Hours — double time | Other rates (if any) | |
|---|---|---|---|---|
| | | | Rate | Hours |
| Monday | | | | |
| Tuesday | | | | |
| Wednesday | | | | |
| Thursday | | | | |
| Friday | | | | |
| Saturday | | | | |
| Sunday | | | | |

4. *Authorized overtime*

Payments to the contractor in respect of authorized overtime shall be governed by the following rules.

(a) No overtime involving extra charge shall be worked without the prior written consent of the Engineer.

(b) *The extra cost* payable to the contractor in respect of authorized overtime shall be only:

    (i) The net labour cost of non-productive hours (as defined in (c) below) at the basic rates shown in Table 2 above (or such revised rates as shall have been published by the authority concerned at the date the overtime is performed).

    (ii) Subject to the approval of the Engineer in each case, such net additional costs and expenses as shall have been necessarily incurred by the contractor as a direct result of the overtime worked, provided such costs and expenses are not recovered or recoverable under any other term or condition of the contract.

(c) 'Non-productive time' shall be the total overtime 'hours' less the actual hours worked. For example a 3-hour period of overtime worked at 'time and a half' is equivalent to 4½ hours paid time. The notional increment of 1½ hours is the non-productive time, paid at the basic rates of Table 2.

(d) The contractor shall provide for the Engineer each week, details in duplicate of all authorized chargeable overtime (on the prescribed overtime sheets). The overtime sheets shall show separately the following data which shall be verified by the signature of the Engineer or his representative:

    The day and date
    Actual hours worked
    Non-productive time alone
    Labour employed, by categories
    Costs calculated in accordance with the above.

5. *Authorized daywork*

(a) The Engineer, when ordering incidental work, may instruct that it be performed at daywork rates: without such specific instructions *in writing* daywork rates shall not be claimable.

(b) For authorized daywork the contractor shall be paid (in lieu of normal payment as prescribed by the contract conditions) as follows:

    (i) Labour at daywork rates defined in (c) below for the hours worked.

    (ii) A fixed percentage thereon as defined in (d) below.

    If overtime is authorized during dayworking, extra payment in respect thereof will be payable to the contractor as specified in Paragraph 4 above.

(c) Daywork rates for the labour categories concerned shall include such of the following items as are applicable in the circumstances of each case:

    Basic labour rates with craftsmen's plus rates.
    Bonus and supplement payments due.
    Daily allowance for travel and subsistence.
    Payments in respect of working conditions.
    Waiting time, including that due to inclement weather.

(d) The percentage referred to in (b) (ii) above shall be ............%.* It shall be deemed to include all the remaining contractor's on-costs, recoveries and profit, such as (but not restricted to) the following:

Standard Rate National Insurance.
Insurance (Contract Works, Third Party, Employers Liability).
Annual and Public Holidays with pay.
Non-contributory sick-pay scheme.
Industrial training levy.
Redundancy payments contribution.
Contract of Employment Act.
All other statutory charges in force at the time.
Site supervision and staff (including foremen and gangers except when working with their gangs when they are paid as workmen).
Small tools.
Protective clothing.
Head office charges.
Profit.

(e) *Materials used on daywork*
The following provisions shall apply only to such materials as are expended on work authorized to be paid at daywork rates and which do not form part of the original design of the works as covered by the contract price.

The cost of materials expended on daywork shall be reimbursed to the contractor at the invoiced price of materials including delivery to site and after deduction of any cash discount in excess of 2½ % and all trade discounts, rebates and allowances.

The contractor is to insert here the percentage addition he will require to be added to the above cost of materials expended on daywork to cover all other charges of every description, including but not limited to, the cost of taking delivery, unloading, unpacking, storing, protecting, hoisting and distributing on site and all establishment charges and profit.

*ADD ............%

(f) *Plant used on daywork*
No payment for use of plant on daywork shall be reimbursable to the contractor unless such plant use has been previously authorized in writing by the Engineer.

The cost of authorized plant used on daywork shall be reimbursed to the contractor at the rates contained in Section 3 of the *Schedules of Dayworks carried out incidental to Contract Works* issued by the Federation of Civil Engineering Contractors and current at the date the work is carried out, or if such schedules shall not be applicable, at such rates as the Engineer shall consider reasonable in the circumstances.

* Tenderers shall insert percentage figures required (as at the date of tender) in each case. For the recognized percentages applicable in (d) to (f) refer to the Federation of Civil Engineering Contractors' publication *Schedules of Dayworks carried out incidental to Contract Works*. The percentages used in determining payments due to the contractor will be those current in the said publication at the date the daywork is performed.

The contractor is to insert here the percentage addition he will require to be added to the above cost of plant expended on daywork to cover all other charges of every description including, but not limited to, establishment charges, profit and any applicable supplementary charges as listed in Section 4 of the above mentioned *Schedule.* ............%.*

TABLE 4

| Type of plant | Hire rate £/hr | Minimum hire period (if any) (hrs) |
|---|---|---|
| | | |
| | | |
| | | |

6. *Authentication*

(Only required when this document is not incorporated in the first paragraph of the Form of Tender — Part 1.)

The data given in this document forms part of our tender number .........................
dated ..................................................... in respect of the following enquiry.

Tender for ..............................................................................
Enquiry reference ..................................... Dated.....................................
        Signed ................................................................................
        Position .................................................................................
        for and on behalf of the Tenderer

Date ................. ................................................................................
................................................................................

## ALTERNATIVE BASIS FOR CALCULATING DAYWORK LABOUR CHARGES

1. Payments for daywork shown in the form of Appendix 11 (above) are frequently based on data provided by the Federation of Civil Engineering Contractors in their publication *Schedules of Daywork carried out incidental to Contract Work* which publication is also named in Clause 52(3) (Dayworks) of the ICE General Conditions of Contract (5th edition). The data has the advantages of being available to and recognized by most employers and engineers and of being updated from time to time as necessary.

2. However, the daywork rate of pay for labour (see Paragraph 5(b) of Appendix 11) includes items some of which are subject to inflation and some to the particular circumstances in which the daywork is performed (e.g. weather, working conditions, etc).

   The Engineer consequently finds it difficult to establish the going rates which will have to be paid to the contractor for any daywork before he authorizes it. Reappraisal of the figures entered at the time of tender may involve a detailed analysis of the contractor's books and vouchers.

3. An alternative basis has therefore found favour with some employers and engineers and has usually been accepted by tenderers and contractors. It involves the transfer out of the daywork labour rates of all ingredients other than the basic labour rate and craftsmen's plus rates. The rest are then included in the contractor's percentage. We are, as a result, concerned with:

   (i) Basic labour rates as set out in Paragraph 2 and Table 2 of Appendix 11. These are revised by national agreements and can be readily established at any given date, from published data.

   (ii) An enhanced contractor's percentage which he enters competitively in Paragraph 5(d) of Appendix 11. It is specified in the contract as being fixed for the whole period of the contract (though the percentage is, of course, applied to the labour charge, which is itself updated to meet inflation and other trends).

4. If this alternative basis is preferred, the forms at Paragraphs 5(b), (c) and (d) of Appendix 11 can readily be revised by the user to suit.

*(Tender for)*

..............................................................................................................................................

*(Enquiry no.)*

..............................................................................................................................................

## FORM OF TENDER – PART ............

## Estimated constructional requirements of electric power on site

The following information is required to enable the Employer and the local electrical supply authority to coordinate site demands for power.

1.* *Three phase supply at ............ V ............ Hz (TP/TP&N/Earth)*

| Contractor's equipment | No. in use | Total connected load (kVA) | Diversity factor | Maximum demand kVA | Contract period of demand |
|---|---|---|---|---|---|
| | | | | | |

Estimated maximum demand (total all 3-phase loads) during contract: ............ kVA

(If variation is large a graph relating estimated maximum demand with contact period should be attached).

2.* *Single phase supply at ............ V ............ Hz.*

| Use | Connected load (kW) |
|---|---|
| Lighting of Works<br>Small power tools<br>Offices (lighting, office machinery)<br>Space heating (offices, stores, etc. | |
| Total connected single-phase load (kW) | |

(Signature only required if this document is not incorporated into the first paragraph of the Form of Tender – Part 1.)

Signed ..................................................................................................

Position ................................................................................................

For and on behalf of ..........................................................................

..............................................................................................................

* To be completed by Employer before issue of form to tenderers.

204

*(Tender for)*

...................................................................................................................................

*(Enquiry no.)*

...................................................................................................................................

## FORM OF TENDER – PART ............

## Schedule of recommended spares

The tenderer is required to list below those spares which he recommends should be purchased to support ............ month's operational use of the plant to be provided under the contract. Use additional sheets if necessary and collect page totals to a price summary sheet at the end.

| Quantity | | Item | Price | |
|---|---|---|---|---|
| Total in plant | Spares recom- mended | | Each | Total (£) |
| | | | | |
| Total price, packed and delivered to Site | | | £ | |

This total price IS/IS NOT* to be included in a lump-sum tender price

(Signature only required if this document is not incorporated into the first paragraph of Form of Tender – Part 1.)

Signed † .................................................................................

Position .................................................................................

For and on behalf of .................................................................

...................................................................................................................................

* Delete as appropriate.
† Where more than one page is used the signature is only required on the price summary sheet at the end.

*(Tender for)*

..........................................................................................................

*(Enquiry no.)*

..........................................................................................................

## FORM OF TENDER — PART ............

### Schedule of prices for use with variations to plant requirements

The following unit prices are to be quoted by the tenderer and may be used for the pricing of variations to the plant and the like. Prices quoted shall be net delivered inclusive of all overheads, expenses, profits or other charges that may be applicable.

| Serial no. | Description and type | Size | Unit price |
|---|---|---|---|
|  |  |  |  |
|  |  |  |  |
|  |  |  |  |
|  |  |  |  |
|  |  |  |  |

(Signature only required if this document is not incorporated into the first paragraph of the Form of Tender — Part 1.)

Signed ..........................................................................................

Position ........................................................................................

For and on behalf of ...................................................................

..........................................................................................................

*Note:* This form is designed for tenders for plant which do not include bills of quantities. For labour rates, a form similar to Appendix 11 Table 2 with paragraph 5(c) can be used.

For constructional works the bills of quantities replace any form such as the above.

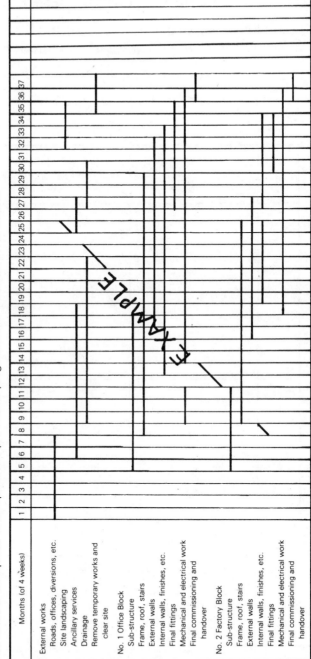

207

## FORM OF TENDER − PART ...........

## Schedule of subcontractors submitted for Engineer's approval (Ref. clause ........... of conditions of contract)

Employer ............................. Tender for .............................

Project ............................... Enquiry No. ...........................

| No. (1) | Portion of the works to be sub-contracted (2) | Name and address of proposed sub-contractor (3) | Status of sub-contractor* (4) |
|---------|-----------------------------------------------|-------------------------------------------------|-------------------------------|
|         |                                               |                                                 |                               |
|         |                                               |                                                 |                               |
|         |                                               |                                                 |                               |
|         |                                               |                                                 |                               |

* For example, Public Limited Company (PLC), Private Limited Company (Co. Ltd), Private Company, Partnership, etc.

*(Tender for)*

................................................................................................................

*(Enquiry no.)*

................................................................................................................

# FORM OF TENDER – PART ............

## Dates for access to the site and contractual times for completion of the works and specified sections thereof

### 1. *Completion*

The tenderer undertakes to execute and complete (including any prescribed tests on completion) the whole of the works and the specified sections thereof in accordance with the times shown in columns 6, 7 and 8 of the schedule attached to this part of the Form of Tender.

In the event of delay in completion of any stated portion or section of the works the contractor shall be liable to pay to the Employer liquidated damages at the rates specified in the contract.

### 2. *Completion of the schedule*

2.1. The Employer's project plan imposes dates before which work on the site cannot be started and by which the sections of the works have to be completed. These dates are shown in the schedule (colums 4 and 5) herewith.

2.2. The tenderer shall insert in the schedule (column 6) the minimum period he will require for producing his drawings and obtaining approvals prior to fabrication, as required by the contract.

2.3. The tenderer shall insert in the schedule (column 7) the minimum period he will require for carrying out work *on site* (to cover erection, installation, commissioning, testing and completion) for each section of the works.

2.4 The tenderer shall insert in the schedule (column 8) his best times for completion of each section of the works based on a contract date on or before ..................

3. If at the time of the enquiry the contract date (i.e. the date on which contract is signed or the date on which a valid instruction to proceed, issued by the Employer is accepted by the contractor, whichever is the earlier) cannot be closely estimated, the dates in the schedule may be expressed in terms of elapsed weeks according to the following rules.

    (a) The week in which the contract date falls is ZERO week. If at that time the Employer has to impose a delay on starting work on the contract, ZERO week shall be the week in which the deferred start date falls.

    (b) The dates for completion shall be deemed to be the SATURDAY of the weeks shown in the schedule (columns 5 and 8).

    (c) The minimum periods required by the contractor shall be deemed to be full weeks (i.e. starting on Monday) – (columns 6, 7 and 8).

4. *Reporting progress of the works*

The contractor shall submit to the Engineer in writing the following data regarding progress of each section of the works.

(a) On Monday of each week following the scheduled date for start of work on site, a report of progress during the previous week in such form as the Engineer shall decide or approve.
(b) On Monday of each week during execution of work on site, a report in duplicate of the labour (by trades) and plant (by types and by horsepower) employed on the work on site during the previous week, showing the approximate hours worked by each main category.
(c) Such additional programmes, reports or data as the Engineer may from time to time require to enable him to establish correctly the state of the works relative to the ᵤcheduled programme.
(d) All progress reports shall specify any obstruction to progress experienced and state the estimated extent of delay resulting, during the period since the last report. Any action taken (or proposed) to overcome the delays shall be included.

*(Tender for)*

.............................................................................................................................................

*(Enquiry no.)*

.............................................................................................................................................

NOTE: Appendix 17, as shown, is typical for a contract in which the contractor is responsible for detailed design of the works and production of the contract drawings. It thus applies chiefly to contracts for plant and machinery, and for 'design and construct' constructional works.

For contracts in which the Employer retains responsibility for design and production of the contract drawings, column 6 of the table, must be placed under 'Employers project dates for' with a new heading 'Issue of construction drawings'. Paragraph 2.2 of the text needs rewording to reflect this change in responsibility, and column 6 must be deleted from paragraphs 1 and 3(c).

## FORM OF TENDER – PART ............

## Schedule of dates for contractor's access to site and for completion of the works and sections thereof

| No. | Location on site | Portions of the works included | Site available to contractor | Completion of the section of the works | Producing drawings for approval | Carrying out all work *on site* | Completing the section of the works |
|---|---|---|---|---|---|---|---|
| | Sections of the works | | Employer's project dates for | | Tenderer to insert minimum period required for | | |
| (1) | (2) | (3) | (4) | (5) | (6) | (7) | (8) |
| 1 | | | | | | | |
| 2 | | | | | | | |
| 3 | | | | | | | |
| 4 | | | | | | | |
| 5 | | | | | | | |
| 6 | | | | | | | |

Tender for ...........................By ........................................

Enquiry No. ...........................Date ....................................

## FORM OF TENDER – PART ............

### Non-compliance with enquiry

The tenderer shall list all matters (technical, commercial or contractual) in which his tender is non-compliant with the requirements of the enquiry.

| Serial no. | Location in enquiry documents (with reference) | Precise matter not complied with | Extent of non-compliance, alternative offered and effect on the works |
|---|---|---|---|
|  |  |  |  |
|  |  |  |  |
|  |  |  |  |
|  |  |  |  |
|  |  |  |  |

Consulting Engineers

# *Truebill, Klinch & Steer*

Ref ..................

Project Management Division
2001 Victoria Street
London SW1   England

Telephone (01) 333–3333
Cables Clincher Westend
Telex 1234567 (Reply back — Trustern)

*(Addressee)*
.............................................................

Date ................

.............................................................

Dear Sirs,

*(Your tender ref.)*
.......................................................................................................

For..............................................................................................................

An examination of your tender dated ............ has disclosed the following apparent error:

| Error location and detail | Incorrect figures | Figures should be |
|---|---|---|
|  |  |  |

Will you please, as a matter of urgency, confirm the existence of this error and let us have your instructions as to the action you wish us to take. In accordance with our rules you may:

*either*   (with lump-sum tenders): maintain the lump-sum price given in your tender and permit rectification of the discrepancy by an equal percentage adjustment to all items in any breakdown of prices given in the tender, or in some other way explained in your reply; (with tenders not based on a lump-sum price): maintain the original rate shown in the bills of quantities of your tender and permit appropriate adjustment of any extensions affected by the error.

*or*   withdraw your tender.

Will you kindly use the enclosed Tender Error Confirmation form for your reply.

Yours faithfully,
for and on behalf of
Truebill, Klinch & Steer

...............................................................
Contracts Engineer

213

# *Truebill, Klinch & Steer*

Your ref. ....................

Our ref. .....................

Project Management Division
2001 Victoria Street
London SW1

Date ........................

Dear Sirs,

## Tender error confirmation

We confirm/do not confirm* the error notified to us in your Tender Error Notification dated ............ Where the error is not confirmed an explanation is given overleaf.

Where the error is confirmed, we elect to adopt the following course.

1. (Where the tender is for a lump-sum price.)*
   To maintain our tender price as originally quoted.
   To rectify the discrepancy caused by the error, we authorize you

   *either:* * To adjust by an equal percentage, all items quoted in the breakdown of prices in our offer;

   *or:* * ...........................................................................................
   ...........................................................................................
   ...........................................................................................

   (Where the tender is based on individual rates.)*
   To maintain the rates quoted in the schedules or bills of quantities in our tender, and adjust the extensions therefrom affected by the error.
2. To withdraw our tender.*

   Signed ...........................................................................

   Position ...........................................................................

   For and on behalf of ...................................................................
   *(tenderer)*
   ...........................................................................................

* Please delete whichever is not applicable as follows:
  in line 1
  alternative course (1) or (2)
  if course (1) is elected, whichever form of tender is inapplicable.
  if lump-sum tender is applicable, whichever alternative is not elected.

Consulting Engineers

# *Truebill, Klinch & Steer*     Ref .................

Project Management Division
2001 Victoria Street
London SW1   England

Telephone (01) 333–3333
Cables Clincher Westend
Telex 1234567 (Reply back — Trustern)

*(Addressee)*
.......................................................     Date .................
.......................................................

Dear Sirs,

*(Employer/Project)*
.......................................................................................
*(Tender for)*
.....................................................  Enquiry No...........

We enclose herewith ........... copies of amendment No. ............, one for each copy of the enquiry documents sent to you on ........*(date)*........ Will you please incorporate the amendments into the documents.

Please detach, complete and return the slip at the bottom of this letter as confirmation that the amendment has been so incorporated, and that its provisions will be taken into account by you when compiling your offer.

Yours faithfully,
for and on behalf of
Truebill, Klinch & Steer

.......................................................
Contracts Engineer

Enclosure ........... copies of amendment No. ............

---

*Receipt slip:*
Please sign, date and return to

# *Truebill, Klinch & Steer*     Reference .........

Project Management Division
2001 Victoria Street,     Date .................
London, SW1

Dear Sirs,

*(Employer/Project)*
.......................................................................................
*(Tender for)*
.....................................................  Enquiry No...........

We have incorporated Amendment No. ........... dated ........... into each of our ...........
copies of the above Tender Documents and will take the contents thereof into account in compiling our offer.

Signed................................................
Date..................................................     For and on behalf of...............................

Consulting Engineers

## *Truebill, Klinch & Steer*

Project Management Division
2001 Victoria Street
London SW1   England

Telephone (01) 333–3333
Cables Clincher Westend
Telex 1234567 (Reply back — Trustern)

## Document transmittal note

To ........................................

From ......................................

Employer ................................................   Contract .......................................

Project .....................................................   Contract ref.....................................

Date.....................

The following documents/drawings originate from ...................................................................
and are forwarded to you for the purpose indicated:

☐ Comment (or 'nil' return) within ...... days     ☐ Copying/Printing/Collating

☐ Approval          ☐ Information          ☐ Return to us

☐ Tender purposes   ☐ Construction        ☐ Correcting/Updating

☐ Other purpose.................................................................................................

| Document/ drawing Ref. No. | Rev. | Title | Number of copies sent | Superseding |
|---|---|---|---|---|
|  |  |  |  |  |
|  |  |  |  |  |
|  |  |  |  |  |
|  |  |  |  |  |
|  |  |  |  |  |
|  |  |  |  |  |

Copies for information to

*Remarks:*

| Addressee | Docs/ Drgs | Tr. note only |
|---|---|---|
|  |  |  |
|  |  |  |
|  |  |  |
|  |  |  |
|  |  |  |
|  |  |  |

Signed......................................

......................................

Consulting Engineers    **STRICTLY CONFIDENTIAL**

Copy No. ......................

## *Truebill, Klinch & Steer*

Project Management Division
2001 Victoria Street
London SW1   England

Telephone (01) 333–3333
Cables Clincher Westend
Telex 1234567 (Reply back – Trustern)

### Record of tenders received

Tenders due on ......... at ......... hrs

Tenders opened on ....... at ....... hrs

Employer...........................................   Contract ............................................

Project............................................   Tenders for ........................................

| Tenderer | Date of tender | Tender price | Remarks |
|---|---|---|---|
| | | | |
| | | | |
| | | | |
| | | | |
| | | | |
| | | | |

Distribution

| Copy No. | Sent to |
|---|---|
| 1 | |
| 2 | |
| 3 | |
| 4 | |
| 5 | |
| 6 | |
| 7 | |

*Tenders opened by*      *Signatures*

1 ........................      ........................

........................

2 ........................      ........................

........................

3 ........................      ........................

........................

Date......................

Consulting Engineers **STRICTLY CONFIDENTIAL**

Copy No. .........

## *Truebill, Klinch & Steer*

Project Management Division
2001 Victoria Street
London SW1   England

Telephone (01) 333–3333
Cables Clincher Westend
Telex 1234567 (Reply back – Trustern)

**Tender appraisal
Part 1: Summary and
recommendations**

Appraisal date ....................

Employer ......................................   Tenders for ..............................,...............

Project ........................................   Enquiry No ....................................

---

1.1   *Distribution*

| Copy No. | To |
|----------|----|
| 1 | |
| 2 | |
| 3 | |
| 4 | |
| 5 | |
| 6 | |
| 7 | |
| 8 | |

1.2 *Enquiry details*

Enquiry ref......................
Issued on ......................
No. of tenders invited ........
No. of tenders received .......
Tender closing date...........
Tenders opened on ...........
Validity expires ................

---

1.3 *Amendments issued during the tender period*

| Ser. No. | Date issued | Effect of amendment |
|----------|-------------|---------------------|
| | | |
| | | |
| | | |
| | | |

---

1.4 *Other notes*

(late/amended tenders; tenderers declining/ignoring enquiry, etc.)

218

1.5 Pre-estimate for Contract £ .................................................................

   (Provisional sums included in estimates £ ..............................................)

   (Contingency included in estimates £ ....................................................)

1.6 *Tenders received*

| | Tenderer | Tender price | |
|---|---|---|---|
| | | As received | Arithmetically corrected |
| 1 | | £ | £ |
| 2 | | | |
| 3 | | | |
| 4 | | | |
| 5 | | | |
| 6 | | | |
| 7 | | | |
| 8 | | | |

1.7 *Alternative offers made by Tenderers*

| Name of Tenderer and details of alternative offered | Effect on tender price | |
|---|---|---|
| | As received | Arithmetically corrected |
| | £ | £ |

1.8 *Tender errors*

Have/have not been notified to Tenderers
Have/Have not been confirmed by Tenderers

---

1.9 *Comparison Summary of Preferred Tenders (not to exceed three)*

---

1.10 *Recommendation*

---

1.11 *Difference between recommended tender price and pre-contract estimate*

---

Date ..........................................    Signed ....................................
Project Manager
(Engineer)

| | |
|---|---|
| Employer ..................................... | **Tender appraisal – Part 2: Précis of tender received** |
| Project ...................................... | |
| Tender for ................................. | |
| Enquiry No. ............................... | (A separate Part 2 is required for each tender received for appraisal) |
| Tenderer .................................... | |
| Tender No. ................................ | |
| Tender date .............................. | Date of appraisal ......................... |

The main features of the tender are:

2.1 *Delivery/Completion dates: Works programme*

2.2 *Comments on additional data provided in enquiry questionnaire*

2.3 *Technical aspects of tender*

2.4 *Contract conditions and regulations*

221

2.5 *Financial terms: payments, credit terms: NPV*

2.6 *Summary of all points of non-compliance with enquiry (including any already mentioned in 2.1–2.5 above)*

2.7 *Matters still requiring negotiation at date of appraisal (and an estimate of effect of decision on appraisal)*

Appraisal by.........................................................................

Date ...................................................................................

# Tender appraisal – Part 3A: Comparison of tender prices

## STRICTLY CONFIDENTIAL

Appraisal by .................................................................   Date.........................   Enquiry No.........................

| Item | Tender 1 ............ A | | Tender 2 ............ B | | Tender 3 ............ C | |
|---|---|---|---|---|---|---|
| | Tender price | Adjustment to common goods/ services | Adjusted tender price | Tender price | Adjustment to common goods/ services | Adjusted tender price | Tender price | Adjustment to common goods/ services | Adjusted tender price |
| | | | | | | |
| Totals | | | | | | |

*Notes:* 1. State reasons for goods/services adjustments..................................................
2. Show deductions in brackets.
3. For more detailed comparison of the financial aspects of the preferred tenderers, see attached report.

223

# Tender appraisal – Part 3B: Discounting tender prices

## STRICTLY CONFIDENTIAL

Tenderer ..... A  Tender No. and Date .....  Enquiry No. ..... TKS 82/4567

| No. (1) | Payments period | (2) Drawings on buyers' credit | (3) Repayment of principal | (4) Interest | (5) Charges (commission commitment management insurance) | (6) Employer's payments in foreign currency (or equivalent) | (7) Total payments in foreign currencies (or equivalent) = (3)+(4)+(5)+(6) | (8) Employer's payments in own currency | (9) Discount factor @ 10% | (10) Foreign currency = (7)×(9) | (11) Own currency = (8)×(9) |
|---|---|---|---|---|---|---|---|---|---|---|---|
| 1 | 1982 – 1 | 0 | 0 | 0 | 21 250 | 990 000 | 1 011 250 | 100 000 | 1.000000 | 1 011 250 | 100 000 |
| 2 | 1982 – 2 | 1 275 000 | 0 | 0 | 21 250 | 75 000 | 96 250 | 200 000 | 0.953463 | 91 771 | 190 693 |
| 3 | 1983 – 1 | 1 700 000 | 0 | 47 813 | 18 062 | 105 000 | 170 875 | 300 000 | 0.909091 | 155 340 | 272 727 |
| 4 | 1983 – 2 | 2 550 000 | 0 | 111 563 | 13 812 | 150 000 | 275 375 | 200 000 | 0.866784 | 238 691 | 173 357 |
| 5 | 1984 – 1 | 1 275 000 | 0 | 207 188 | 7 437 | 75 000 | 289 625 | 150 000 | 0.826446 | 239 359 | 123 967 |
| 6 | 1984 – 2 | 1 700 000 | 0 | 255 000 | 4 250 | 105 000 | 364 250 | 50 000 | 0.787986 | 287 024 | 39 399 |
| 7 | 1985 – 1 | | 708 333 | 318 750 | | | 1 027 083 | | 0.751315 | 771 663 | |
| 8 | 1985 – 2 | | 708 333 | 292 187 | | | 1 000 521 | | 0.716351 | 716 724 | |
| 9 | 1986 – 1 | | 708 333 | 265 625 | | | 973 958 | | 0.683014 | 665 227 | |
| 10 | 1986 – 2 | | 708 333 | 239 063 | | | 947 396 | | 0.651228 | 616 970 | |
| 11 | 1987 – 1 | | 708 333 | 212 500 | | | 920 833 | | 0.620921 | 571 765 | |
| 12 | 1987 – 2 | | 708 333 | 185 938 | | | 894 271 | | 0.592026 | 529 432 | |
| 13 | 1988 – 1 | | 708 333 | 159 375 | | | 867 708 | | 0.564474 | 489 799 | |
| 14 | 1988 – 2 | | 708 333 | 132 813 | | | 841 146 | | 0.538205 | 452 709 | |
| 15 | 1989 – 1 | | 708 333 | 106 250 | | | 814 583 | | 0.513158 | 418 010 | |
| 16 | 1989 – 2 | | 708 333 | 79 688 | | | 788 021 | | 0.489277 | 385 560 | |
| 17 | 1990 – 1 | | 708 333 | 53 125 | | | 761 458 | | 0.466507 | 355 226 | |
| 18 | 1990 – 2 | | 708 333 | 26 563 | | | 734 896 | | 0.444798 | 326 880 | |
| Totals | | 8 500 000 = 8 500 000 | 8 500 000 | 2 693 437 | 86 062 | 1 500 000 | 12 779 500 | 1 000 000 | | 8 323 400 | 900 143 (12) |

Conversion foreign to home at exchange rate £1.00 = $US2.00 (if not converted before entry on this sheet)

8 323 400 = 4 161 700 (13)

Tender NPV (12)+(13) = 5 061 843

Appraisal made by .....

Date .....

Use separate sheet for each currency and total NPVs (or convert and work in own currency).

**Foreign currencies in tender price**

| Currency | Conversion rate = | Total payments in tender price | NPV of total payments |
|---|---|---|---|
| 1 $US | £1 = 2.000 | 12 779 500 | 8 323 401 |
| 2 | | | |
| 3 | | | |
| 4 | | | |
| | | 12 779 500 | 8 323 401 |

# Tender appraisal – Part 4: Technical comparisons – preferred tenderers

*Table 4A: Comparison of plant and services offered*

Enquiry No. ..............  Tender for ..............

| Serial No. | Plant and services | Tenderer 'A' | Tenderer 'B' | Tenderer 'C' |
|---|---|---|---|---|
| | List items to be compared<br>Include, as applicable:<br>  Works testing and proving<br>  Packing and transport<br>  Loading and unloading<br>  Storage and insurance<br>  Erection at site<br>  Commissioning<br>  Documentation<br>  Operator training<br>  Cabling and protection<br>  Painting and finishing<br>  Weights of main items<br>  Spare parts<br>  Delivery period<br>  Special tools<br>  Lubricants (first fill) | | | |

*Notes:* 1. Include sufficient detail for each item to facilitate comparisons.
2. If no details are available, show item as 'included' or 'not included'.
3. If data obtained subsequent to tender, give detail of letter, telex, etc.
4. Show no prices on this table.

**Tender appraisal – Part 4: Technical comparisons – preferred tenderers**

*Table 4B: Comparison of performances of plant offered*

Enquiry no............... Tender for ..................

| Serial No. | Performance | Tenderer 'A' | Tenderer 'B' | Tenderer 'C' |
|---|---|---|---|---|
| 1 | Number of machines | | | |
| 2 | Performance (output and accuracy) up to designed overload. Detail items. | | | |
| 3 | Maximum production rates (at full load) of main products. | | | |
| 4 | Recommended down-time for maintenance; operating hours per year for scheduled annual output of each product. | | | |

# Tender appraisal – Part 4: Technical comparisons – preferred tenderers

*Table 4C: Comparison of energy and service requirements*

Enquiry no. ........................ Tender for ........................

| Serial No. | Service | Tenderer 'A' | Tenderer 'B' | Tenderer 'C' |
|---|---|---|---|---|
| 1 | *Power demand*<br>Total installed power (kW)<br>Normal running load (kVA)<br>Peak running load (kVA)<br>Peak starting load (kVA)<br>Duration of peak (secs)<br>Motors, types and ratings | | | |
| 2 | *Energy consumption*<br>Running light<br>¼-production rate*<br>½-production rate*<br>Full production rate*<br>Designed overload* | | | |
| 3 | *Other services*<br>(Consumption of gas, water, compressed air, process gases, etc as applicable) | | | |

* i.e. kWh per tonne of product (in all cases)

227

(Sent on Employer's headed letter-paper)

................................................ Reference ...................................

*(Tenderer's name*

................................................

*and address)*

................................................ Date ...........................................

Dear Sirs,

*(Project)*

................................................................................................

Tender No. ............... for ....................................................................

*(contract title)*

1. We are pleased to inform you that, subject to a satisfactory outcome to negotiations between us on certain points in your tender as above, the contract for this work will be placed with your company.

    To allow work to be put in hand without delay will you kindly accept this letter as an INSTRUCTION TO PROCEED (subject to the terms and conditions of contract other than those in dispute between us) with the following parts of the contract works forthwith:

    (a) ........................................................................................

    (b) ........................................................................................

    (c) ........................................................................................

2. This Instruction is subject to an over-riding limit of total expenditure (incurred and committed) of a contract value of £............ which figure shall not be exceeded witout prior written authority from me.

3. On placing the full contract with you all liabilities and benefits of both parties will be absorbed thereby and this Instruction to Proceed shall be determined. In the event the aforementioned negotiations fail to lead to agreement so that a full contract cannot be concluded between us we shall have the right to determine by notice in writing the contract formed from this Letter of Instruction, provided always that you shall be paid in such event the contract value of all portions of the Works properly executed thereunder together with any other costs and expenses necessarily incurred by you as a result of the prior determination. Property in all materials acquired or work carried out shall forthwith vest in us (if not already so vested) and be held available by you at our disposal.

4. Will you kindly let me have by first-class return of post your unconditional acceptance of this Instruction to Proceed, and thereupon put the work in hand without delay.

    Yours faithfully,

*(Position)*

................................................................................................

for and on behalf of

*(Employer)*

................................................................................................

## LETTERS OF ACCEPTANCE

The formal acceptance of the chosen tender and the setting up of a contract must be done by the Employer. The Engineer's role is one of advice, to ensure that a proper and unambiguous relationship is duly established. This may include the preparation of any necessary documents in the Employer's name.

Three typical Letters of Acceptance are given below: they are not standard forms, but need drafting to suit each contract.

(a) Acceptance of a tender (which has not been modified by negotiation) for supply and erection of Plant.

(b) Acceptance of a tender (which has not been modified by negotiation) for civil engineering works.

(c) Placing a contract based on a tender modified by negotiation. In practice this document will usually not form a legal 'acceptance' but will be a counter offer, itself needing acceptance.

It is sometimes helpful to include a number of other subjects in a letter of acceptance, such as those which:

- Need immediate action.
- Relate to early stages of the contract.
- Are of sufficient importance to be confirmed or stressed (even though they already appear in the enquiry or tender).

Note that if any such matters can be held to be new terms or conditions, they will convert an unconditional acceptance into a counter-offer ((c) above). Possible subjects might include, for example, the following. Their inclusion must never be allowed to fog the main object of the letter; if they are numerous or involved, write separate instructions about them.

- Payments due to be made 'with order'.
- Letters of Credit to be established by the Employer: name and address of Contractor's bank.
- Names, addresses and name of site representative of any quantity surveyors appointed for the contract.
- Request for names of contractors site management and their management structure.
- The requirement for the contractor to submit insurance policies and receipts, as in the contract.
- The need for the contractor to produce a performance bond to the Engineer, if required by the contract.
- Safety precautions – site regulations.
- Security measures, (a) of persons (b) of site.
- Initial co-ordination meeting – place and time.
- Employer's import agents (overseas contracts).

(Sent on Employer's headed letter-paper)

..............................................  Reference ...................................
*(Tenderer's name*
....................................................
*and address)*
..............................................  Date ...........................................

Dear Sirs,

*(Project)*
.........................................................................................................
*(contract title)*
Tender No. ............... for ..............................................................

1. Your tender as above dated ............................................ for the supply and erection of the above plant is hereby accepted and together with this letter shall form a binding contract between us pending the signing of a formal Agreement as provided in the conditions of contract.
2. The Contract Price is confirmed at £............ which is subject to contract price adjustment in accordance with the terms of payment, the base date being ..........
.........................................................................................................
3. We have appointed as Engineer for the Contract ..........................................
of ....................................................................................................
Telephone ...................... Telex ...................... Cables ......................
The Engineer's Representative shall be ........................................................;
the duties delegated to him are being drawn up and will be notified to you shortly.
4. The Contract Date is ...................................... You are reminded that delivery of the Plant to site is due on or before ........................................................
and completion of the Works by ...............................................................
5. You are instructed to get in touch with the Engineer without delay to arrange further details regarding execution of the Contract. You should send to him for approval as soon as possible the policies of insurance specified in the contract conditions.
6. Kindly acknowledge receipt of this Letter of Acceptance.

Yours faithfully,

*(Position)*
.........................................................................................................
for and on behalf of
*(Employer)*
.........................................................................................................

*Letter of Acceptance of a tender for civil works (unmodified)*

(Sent on Employer's headed letter-paper)

..............................................
*(Tenderer's name*
..............................................
*and address)*

Reference ...................................

Date ...........................................

Dear Sirs,

.................................................................................................................
*(Project)*
Tender No. ............... for ..................................................................
*(contract title)*

1. Your tender as above dated ........................................ for the construction, completion and maintenance of the above Works is hereby accepted and together with this letter shall form a binding contract between us pending the signing of a formal Agreement as provided in the conditions of contract.

2. The Works shall be priced in accordance with the prices, rates, terms and conditions in your tender and are subject to remeasurement of quantities. The prices and rates are to be subject to cost price variation as provided in clause ............ of the conditions.

3. We have appointed as Engineer for the Contract ..........................................
of .......................................................................................................
Telephone ...................... Telex ...................... Cables ......................
The Engineer's Representative on site shall be ..............................................
and the Quantity Surveyor appointed by the Engineer is to be Mr ......................
The duties delegated to these officials are being drawn up and will be notified to you shortly.

4. The Contract Date is ..................................................... and the site will be available to you from a start date of ..................................................... You are reminded that the achievement of progress on the Works and completion in accordance with the programme included in the tender is of first importance to ensure coordination with other contracts in progress at the site.

5. You will be required as provided in clause ............ of the conditions of contract to negotiate contracts with the undermentioned nominated sub-contractors on the basis of arrangements we have made with them. Copies of documents relating to these sub-contracts will be sent to you separately to enable you to do this. Prime Cost Sums are included in the bills of quantities in respect of these sub-contracts.

With .............................................. of ..............................................

for ..................................................................... at a cost of £............

231

6. You are instructed to get in touch with the Engineer without delay to arrange further details regarding the performance of the contract. You should send him for approval, before starting work on site, the policies of insurance specified in the contract conditions.
7. Will you kindly acknowledge receipt of this Letter of Acceptance.

Yours faithfully,


*(Position)*
........................................................................................................
for and on behalf of
*(Employer)*
........................................................................................................

# APPENDIX 26C  *Award of a contract based on a tender for civil works (modified by negotiation)*

(Sent on Employer's headed letter-paper)

..................................................  Reference ....................................

, *(Tenderer's name*

..................................................

*and address)*

..................................................  Date ........................................

Dear Sirs,

*(Project)*

..........................................................................................................

Tender No. ............... for ................................................ *(contract title)* ...........

1. Following our recent discussions, it has now been decided to place with you the above contract on the basis of your tender dated ..........................................
   extended and amended in the manner agreed during our negotiations and recorded in

   (a)  (List in date order all letters, minutes, reports, etc. setting out the changes to
   (b)  the tender or recording agreement on them.)
   (c)
   (d)
   etc.

   These documents, your tender, this letter and your acceptance hereof constitute the Contract Documents, which will form a binding contract between us pending completion of a formal confirming Agreement.

2. The Works will be priced at the prices, rates, terms and conditions in your tender and are subject to remeasurement of quantities. Prices and rates are to be subject to cost price variation as provided in clause ............ of the conditions.

3. We nominate for appointment as Engineer for the Contract ............................
   of ......................................................................................................
   Telephone ...................................... Telex .......................................
   The Engineer's Representative will be .........................................................
   assisted by a quantity surveyor ................................................................
   The duties delegated to these officials are being drawn up and will be notified to you on concluding the contract.

4. We intend nominating the following sub-contractors with whom you will be required to enter nominated sub-contracts as provided in clause ............ of the conditions. Prime Cost Sums are included in the bills of quantities in respect thereof.

   With .............................................. of ............................................

   for ........................................ at a cost of £........................................

233

5. Will you please let us have as soon as possible your formal Letter of Acceptance of the contract as proposed and described in this letter, signed by a duly authorized official of your company.

6. The date of your Letter of Acceptance will be the contract date and the site will be available to you for start of work ............ working days thereafter. You are reminded that the achievement of progress on and completion of the Works in accordance with the programme included in the tender is of first importance to ensure coordination with other contracts in progress at the site.

7. Following acceptance of the contract you should contact the Engineer without delay to arrange further details regarding its performance. You should also send him for approval, before starting work on site, the policies of insurance specified in the contract conditions.

Yours faithfully,


*(Position)*
.........................................................................................
for and on behalf of

*(Employer)*
.........................................................................................

Consulting Engineers

# *Truebill, Klinch & Steer*                         Ref ...................

Project Management Division
2001 Victoria Street
London SW1  England

Telephone (01) 333–3333
Cables Clincher Westend
Telex 1234567 (Reply back — Trustern)


*(Addressee)*
.................................................................                Date .................
.................................................................


Dear Sirs,

  Employer/Project........................................................................
  Contract.............................................................Enquiry No..........

With reference to the Tender No. .......... which you submitted on .......... in connection with the above named enquiry, we regret to inform you that your offer has not been successful.

If you have not already done so, would you kindly return to us, marked for my attention, all documents issued to you for the purposes of this enquiry.

On behalf of our Client we thank you for the care and interest you have taken in preparing and presenting your offer, and assure you that the decision in this instance will in no way adversely affect any offers you may be able to make for work of a similar nature in future.


Yours faithfully,
for and on behalf of
Truebill, Klinch & Steer


.................................................................
Contracts Engineer

## Performance bond

BY THIS BOND made the ................ day of ................ one thousand nine hundred and ................
We ...................................................................................................................
(hereinafter called 'the Contractor') and .......................................... whose head office is situate at ..................................... (hereinafter called 'the Sureties') are held and firmly bound unto ...........................................................................................
(hereinafter called 'the Employer')
so that

WHEREAS by an Agreement (hereinafter called 'the Agreement') dated ................
................ the Contractor and the Employer entered into an agreement as therein stated.

NOW WE the Sureties hereby jointly and severally guarantee to the Employer punctual true and faithful performance and observance by the Contractor of the covenant on the part of the Contractor contained in the Agreement and undertake to be responsible to the Employer his legal personal representatives, successors or assigns as Sureties for the Contractor for the payment by the Contractor of all sums of money losses damages costs charges and expenses that may become due or payable to the Employer his legal personal representatives successors or assigns by or from the Contractor by reason or in consequence of the default of the Contractor in the performance or observance of the said covenant on the part of the Contractor but so nevertheless that the total amount to be demanded or recovered by the Employer his legal representatives successors or assigns of or from us as Sureties shall not exceed ..................................................................................................................

This Guarantee shall not be revocable by notice and our liability as Sureties hereunder shall not be impaired or discharged by any extensions of time or variations or alterations made, given conceded or agreed (with or without our knowledge or consent) under the terms and conditions contained in the Agreement or (where the Employer or the Contractor is a Firm) by a change in the constitution of the Employer's or the Contractor's respective firms.

Any demands for payment under this Guarantee shall be made upon us in writing and such payment by us shall be made without objection on our being given evidence as to the existence of a default by the Contractor and of the damages due or payable to the Employer in respect thereof which shall be a certificate of the findings of a court of law or of arbitration (in cases in which the default has been the subject of proceedings therein) or otherwise the certificate of the person named as the Engineer in the Agreement aforesaid.

236

In witness whereof the Contractor and the Sureties have caused their respective Common Seals to be hereunto affixed the day and year first before written.

The Common Seal of.......................
...............................................
was hereunto affixed in the presence of
...............................................
...............................................
...............................................

The Common Seal of .....................
...............................................
was hereunto affixed in the presence of
...............................................
...............................................
...............................................

## Performance bond

BY THIS BOND made the ...................... day of ...................... one thousand nine hundred and ........................

We ............................................................................................................

whose registered office is situate at ..............................................................

.......................................................... (hereinafter called 'the Guarantors')

are held and firmly bound unto ................................................................ of

.......................................................... (hereinafter called 'the Employer')

so that

WHEREAS the Guarantors have a controlling interest in ....................................

.......................................................................... whose registered office is

at ........................................................................ (hereinafter called 'the

Contractor') and whereas the Employer has entered (or is about to enter) a contract or contracts with the Contractor for the execution of certain works required by the Employer namely .......................................................................................

NOW WE the Guarantors hereby guarantee to and covenant with the Employer that in the event that the Employer shall have duly entered a contract or contracts as aforesaid with the Contractor then the Contractor will truly punctually and faithfully perform and observe all the obligations terms provisions conditions and stipulations in the said contract or contracts mentioned or described or to be implied therefrom on its part to be so performed and observed according to the true purport intent and meaning thereof and if for any reason whatsoever and in any way the Contractor shall fail to perform and observe the same then the Guarantors shall take over from the Contractor and shall forthwith perform and observe or cause to be performed and observed such obligations terms provisions conditions and stipulations as aforesaid and shall be responsible as Sureties for the Contractor to the Employer his legal personal representatives successors or assigns for the payment by the Contractor of all sums of money losses damages costs charges and expenses that may become due or payable to the Employer his legal personal representatives successors or assigns by reason or in consequence of the acts or defaults of the Contractor in the performance or observance of the obligations terms provisions conditions and stipulations aforesaid.

The Guarantors shall not in any way be released from liability hereunder by any extension of time or variations or alterations made given conceded or agreed (with or without their knowledge or consent) under the terms and conditions contained in the contract or contracts or in the extent or nature of the works to be constructed completed or maintained thereunder or by any allowance of time or forbearance or forgiveness in or in respect of any matter or thing concerning the said contract or contracts as the case may be or by a change in the constitutions of the Employer's or Contractor's respective firms.

This Bond shall be revocable at the option of the Guarantors at any time as to any future contracts that might be placed by the Employer with the Contractor by one months notice in writing given to the Employer by the Guarantors but without prejudice to the rights of the Employer his legal personal representatives successors or assigns hereunder in respect of any contract or contracts that may have been placed by the Employer with the Contractor prior to the date on which any such notice expires.

In witness whereof the Guarantors have caused their Common Seal to be hereunto affixed the day and year first before written.

The Common Seal of......................

................................................

................................................

was hereunto affixed in the presence of

................................................

................................................

THIS AGREEMENT made the .......................... day of .......................... one thousand nine hundred and ........... between ................................................ of ....................................................................................................................... ('the Employer') of the one part and ............................................................. of ....................................................... ('the Sub-Contractor') of the other part witnesseth that

WHEREAS the Employer intends entering a contract ('the Main Contract') with a contractor ('the Main Contractor') for the construction completion and maintenance of ......................................................................................................................... ('the Works') subject to terms and conditions generally in accordance with

*(here indicate the General Conditions, Special Conditions,*
.............................................................................................................................

*Contract Programme, Liquidated Damages, Site Regulations*
.............................................................................................................................

*which are expected to apply)*
.............................................................................................................................

and whereas the Sub-Contractor has at the invitation of the Employer tendered on
....................................................... for .......................................................
('the Sub-Contract Works') forming part of the Works as a sub-contractor nominated (or to be nominated) by the Employer in the Main Contract (the said tender being hereinafter called 'the Offer') it is hereby agreed between the parties hereto as follows:

Art. 1 In consideration of the Employer having invited the Sub-Contractor to tender for the Sub-Contract Works and having agreed on the basis of the Offer to nominate the Sub-Contractor to the Main Contractor as a nominated sub-contractor under the Main Contract the Sub-Contractor undertakes

(a) To transfer the Offer in the same terms and conditions mutatis mutandis to the Main Contractor on his appointment by the Employer

(b) As and when required by the Main Contractor to enter negotiations with the Main Contractor in good faith on the basis of the Offer for a sub-contract as a nominated sub-contractor under the Main Contract.

Art. 2 This Agreement shall come into effect on the date first above mentioned and shall remain in force for a period of ........... calendar months thereafter unless extended by the parties in writing or unless it shall have been previously determined by the Employer in accordance with Art. 3.

Art. 3 The Employer shall be entitled to determine this Agreement by seven days notice to the Sub-Contractor in writing without liability to either party if

either

(a) The Main Contractor after appointment shall in the exercise of his rights under the terms and conditions of the Main Contract validly object to the appointment of the Sub-Contractor as a nominated sub-contractor

or

(b) The Employer shall for any reason whatsoever elect not to undertake such part or parts of the Works as include the Sub-Contract Works or

(c) Following negotiations in paragraph (b) of Article 1 aforementioned the Employer shall be of the opinion that no valid sub-contract is likely to be agreed within a reasonable period.

Art. 4 Nothing contained in the Offer or in any modifications or additions thereto negotiated and agreed between the Sub-Contractor and the Main Contractor shall operate to limit the Sub-Contractor's obligations under this Agreement.

IN WITNESS WHEREOF the parties hereto have set their hands by the duly authorized representatives the date first above mentioned.

Signed for ....................................  Signed for ....................................

..............................................  ..............................................

by ...........................................  by ...........................................

Position .....................................  Position .....................................

Witness ......................................  Witness ......................................

Occupation....................................  Occupation....................................

# *Truebill, Klinch & Steer*

Project Management Division     Date . . . . . . . . . . . . . . . . . . . . . . . . . . . . . . . . . . . . . . . . . . . . .

## Notice of placing of contract (plant)

Employer . . . . . . . . . . . . . . . . . . . . . . . .     Site . . . . . . . . . . . . . . . . . . . . . . . . . . . . .

Project . . . . . . . . . . . . . . . . . . . . . . . . . . .     Contract No. . . . . . . . . . . . . . . . . . . . . . . . .

For . . . . . . . . . . . . . . . . . . . . . . . . . . . . . . . . . . . . . . . . . . . . . . . . . . . . . . . . . . . . . . . . . . . . .

1. Please note that on . . . . . . . . *(date)* . . . . . . . . the Employer concluded a contract as above
    with . . . . . . . . . . . . . . . . . . . . . . . . . . . . . . . . . . . . . . . . . . . . . . . . . . . . . . . . . . . . . . . . . . . . . . . . .
    of . . . . . . . . . . . . . . . . . . . . . . . . . . . . . . . . . . . . . . . . . . . . . . . . . . . . . . . . . . . . . . . . . . . . . . . . . . . .
    Telephone No . . . . . . . . . . . . . . . . . . . . . . . . . . . . . . . Telex . . . . . . . . . . . . . . . . . . . . . . . . . . . . . . .
    Plant will be assembled by contractor at . . . . . . . . . . . . . . . . . . . . . . . . . . . . . . . . . . . . . . . . . . .
    The contractor's liaison engineer is Mr . . . . . . . . . . . . . . . . . . . . . . . . (Telephone . . . . . . . . . . . .)
    The contractor's site representative will be Mr . . . . . . . . . . . . . . . . . (Telephone . . . . . . . . . . . .)

2. Our Project Manager is Mr . . . . . . . . . . . . . . . . . . . . . . . . . . . . . . . . . . . . . . (extn . . . . . . . . . . . .)
    Our Project Engineer concerned is Mr . . . . . . . . . . . . . . . . . . . . . . . . . . . . . . . (extn . . . . . . . . . . . .)
    The Engineer's Representative on site will be Mr . . . . . . . . . . . . . . . . . . . . . . . . . . . . . . . and his
    telephone on site will be . . . . . . . . . . . .

3. (a) The contract includes supply/shipment/erection/supervision of erection/
        completion/defects liability for . . . . . . . . . . months after take-over/Maintenance.
    (b) Delivery is ex works/FOR/FOB/C & F/CIF/to site by . . . . . . . . *(date)* . . . . . . . .
    (c) Completion to take-over is due by . . . . . . . . *(date)* . . . . . . . .
    (d) Risk in plant passes to Employer on . . . . . . . . . . . . . . . . . . . . . . . . . . . . . . . . . . . . . . . . . . . . .
    (e) Contract Price is fixed/subject to CPA by . . . . . . . . . . . . formula
    (f) Terms include/do not include an initial 'with order' payment
    (g) The contract includes/does not include Prime Cost Sums/Provisional Sums
        as detailed in Paragraph 6 below
    (h) Standard Form of Conditions applicable is . . . . . . . . . . . . with/without amend-
        ments. There are/are not additional Special Conditions, which include:

4. The Contract specifies formal inspections/demonstrations/tests.

  — At contractor's factory prior to despatch.
  — On delivery to site.
  — On completion of erection ('take-over') on site.
  — Guaranteed performance tests to take place before ........*(date)*........
  — ..........................................................................................................

5. The contractor has to produce for the Engineer's approval
  (a) by ........*(date)*........ a Performance Bond by a bank/insurance company
  (b) evidence of suitable insurances held by him in respect of the Contract Works/
      Public liability/Employer's liability/Plant in transit/...................................
      ............................................... / ...............................................

6. *Other comments:*

Extn No. ...........................................    ...................................................
Date ...........................................    Contracts Engineer

Distribution
(with copy No. if a copy of the Contract is enclosed)

| To | Copy No. | To | Copy No. |
|---|---|---|---|
|  |  |  |  |
|  |  |  |  |

# Truebill, Klinch & Steer

Project Management Division      Date .......................................................

## Notice of placing of contract (construction)

Employer .....................................    Site ......................................

Project .......................................    Contract No. ...............................

For ..............................................................................................

1. Please note that on ........*(date)*........ the Employer concluded a contract as above
   with ......................................................................................
   of .........................................................................................
   Telephone No. ........................................ Telex ................................
   The contractor's liaison engineer is Mr ......................... (Telephone ............)
   The contractor's representative on site is Mr ................... (Telephone ............)

2. Our Project Manager is Mr ................................................. (extn ............)
   Our Project Engineer concerned is Mr ..................................... (extn ............)
   Our Engineer's Representative is Mr ................................................. and his
   telephone number on site will be .........................................................
   The Quantity Surveyors are ...............................................................
   of ................................................. Telephone ............ Telex ............

3. (a) The contract is payable Lump Sum/On firm Bills of Quantities/on remeasure-
       ment/Interim Certificates/monthly/........................./.........................
   (b) The Start Date is .....................................................................
   (c) Completion Date is ....................................................................
   (d) There are/are not contractual intermediate dates (See Paragraph 6 below)
   (e) Contract Prices are fixed/subject to CPA by ............ formula
   (f) Minimum interim certificate value £............
   (g) Bills of Quantities include Prime Cost/Provisional sums. (See Paragraphs 5
       and 6 below for details.)
   (h) The period of maintenance of the Works is ............ months after the date of
       completion.
   (j) Standard Method Measurement used is .................................................
   (k) Standard Form of Conditions of Contract is...........................................
       with/without amendments. There are/are not additional Special Conditions,
       which include:

4. The contractor has to produce for the Engineer's approval:
   (a) by ........*(date)*........ a Performance Bond by a bank/insurance company
   (b) evidence of suitable insurances held by him in respect of the Contract Works/Public liability/Employer's liability

5. The following are nominated as sub-contractors for the sub-contract works shown

| Nominated Sub-contractor | Sub-contract for | Prime cost sum in BOQ |
|---|---|---|
|  |  |  |

6. *Other comments:*
   (Include Special Conditions, Provisional Sums, etc.)

Extn No. .......................................   ..................................................
Date ...........................................   Contracts Engineer

Distribution
(with copy No. of any set of contract documents attached)

| To | Copy No. | To | Copy No. |
|---|---|---|---|
|  |  |  |  |

# Index

247